Fundamental Aspects of Crystallization and Precipitation Processes

Daina M. Briedis and Kuttanchery A. Ramanarayanan, editors

A. Abdul-Rahman	R.W. Farmer	G.H. Nancollas
J.R. Beckman	J. Garandet	H. Narayanan
G.D. Botsaris	M.E. Glicksman	J.O. Pagounes
J. Budz	E. Halter	R.W. Peters
C.T. Chang	S.M. Hamza	A.D. Randolph
T.K. Chang	J.P. Hsu	R.W. Rousseau
P.P. Chiang	A.G. Jones	M. Saska
S.T. Chou	J. Kanel	D. Schruben
M.D. Donohue	P.H. Karpinski	S. Tsurouka
M.A. Farrell Epstein	W.F. Klima	G.R. Youngquist
L.T. Fan	S.P. Marsh	R.C. Zumstein
	J.W. Mullin	

AIChE Symposium Series

1987

Number 253 — Volume 83

Published by
American Institute of Chemical Engineers

345 East 47 Street — New York, New York 10017

Copyright 1987

American Institute of Chemical Engineers
345 East 47 Street, New York, N.Y. 10017

AIChE shall not be responsible for statements or opinions advanced in papers or printed in its publications.

Library of Congress Cataloging-in-Publication Data
Fundamental aspects of crystallization and precipitation processes.

(AIChE symposium series ; no. 253, v. 83)
Includes index.
1. Crystallization—Congresses. 2. Precipitation (Chemistry)—Congresses.
I. Bredis, Diana M., 1956-   . II. Ramanarayanan, Kuttanchery A.,
1952-   . III. Series: AIChE symposium series ; no. 253.
QD548.F86 1987   548'.5   87-952
ISBN 0-8169-0425-1

Authorization to photocopy items for internal or personal use, or the internal or personal use of specific clients, is granted by AIChE for libraries and other users registered with the Copyright Clearance Center (CCC) Transactional Reporting Service, provided that the $2.00 fee per copy is paid directly to CCC, 21 Congress St., Salem, MA 01970. This consent does not extend to copying for general distribution, for advertising or promotional purposes, for inclusion in a publication, or for resale.

Articles published before 1978 are subject to the same copyright conditions and the fee is $2.00 for each article. AIChE Symposium Series fee code: 0065-8812/87 $2.00

Printed in the United States of America by
Twin Production & Design

QD
548
.F86
1987

FOREWORD

This symposium series is a compilation of papers presented at the National AIChE meetings in Chicago (1985) and New Orleans (1986). More than 30 papers were presented in 4 sessions. A broad range of topics are covered by the seventeen papers in this issue. Topics include macroscopic properties of solutions; nucleation, growth and agglomeration kinetics and dynamics and modeling of crystallizers. These papers have provided us with a much better understanding of crystallization and precipitation processes and illustrate the continued high level of activity in this field of research.

Daina M. Briedis
Department of Chemical Engineering
Michigan State University
East Lancing, MI

Kuttanchery A. Ramanarayanan
Chemical Engineering Development
Hoffmann-La Roche Inc.
Nutley, NJ

CONTENTS

FOREWORD iii

SOME PROPERTIES OF SUPERSATURATED SOLUTIONS .... H. Narayanan and G. R. Youngquist   1

CRYSTALLIZATION FROM IONIC SOLUTION .... P. P. Chiang and M. D. Donohue   8

EFFECTS OF IMPURITIES ON THE PRODUCTION AND SURVIVAL STEPS OF CONTACT NUCLEATION
.... G. D. Botsaris and J.O. Pagounes   19

THE GROWTH OF GYPSUM .... W. F. Klima and G. H. Nancollas   23

WAX HABIT IN CHILLED DIESEL FUEL, A GROWTH MODEL .... E. Halter, D. Schruben and J. Kanel   31

THE GROWTH AND DISSOLUTION OF STRONTIUM FLUORIUDE. INFLUENCE OF INHIBITORS
.... A. Abdul-Rahman, S. M. Hamza and G. H. Nancollas   36

KINETICS OF SUCROSE CRYSTALLIZATION FROM IMPURE SOLUTIONS .... M. Saska and J. Garandet   42

SOLUTION GROWTH UNDER NON-UNIFORM CONDITIONS—ADP AND $MgSO_4.7H_2O$
.... H. Narayanan and G. R. Youngquist   47

DIRECT USE OF THE BCF CRYSTAL GROWTH MODEL IN CORRELATING EXPERIMENTAL DATA
.... P. H. Karpinski   54

THE EFFECT OF Pb(II) AS A TRACE IMPURITY ON THE CRYSTALLIZATION KINETICS OF $CaCO_3$ PRECIPITATION
.... R. W. Peters and T. K. Chang   62

AGGLOMERATION OF POTASSIUM SULFATE CRYSTALS IN AN MSMPR CRYSTALLIZER
.... J. Budz, A. G. Jones and J. W. Mullin   78

BIMODAL CSD BARITE DUE TO AGGLOMERATION IN AN MSMPR
   CRYSTALLIZER .... J. R. Beckman and R. W. Farmer   85

CRYSTALLITE AGING UNDER TRANSPORT LIMITED CONDITIONS—APPLICATIONS OF MULTIPARTICLE
   DIFFUSION ALGORITHMS .... S. P. Marsh and M. E. Glicksman   95

STATE SPACE REPRESENTATION OF THE DYNAMIC CRYSTALLIZER POPULATION BALANCE: APPLICATION
   TO CSD CONTROLLER DESIGN .... S. Tsuruoka and A. D. Randolph   104

SIMULATION STUDIES OF A FEEDBACK CONTROL STRATEGY FOR BATCH CRYSTALLIZERS
.... C. T. Chang and M. A. Farrell Epstein   110

TRANSIENT ANALYSIS OF CRYSTALLIZATION: EFFECT OF THE SIZE-DEPENDENT RESIDENCE TIME OR
   CLASSIFIED PRODUCT REMOVAL .... L. T. Fan, S. T. Chou and J. P. Hsu   120

UTILIZATION OF INDUSTRIAL DATA IN THE DEVELOPMENT OF A MODEL FOR CRYSTALLIZER SIMULATION
.... R. C. Zumstein and R. W. Rousseau   130

SOME PROPERTIES OF SUPERSATURATED SOLUTIONS

H. Narayanan ■ Clarkson University, Potsdam, NY 13676
G. R. Youngquist ■ Iowa State University, Ames, IA 50011

Density, viscosity, and electrical conductivity have been measured for undersaturated and supersaturated aqueous solutions of several inorganic salts. In contrast to recent results for diffusivities, no unusual behavior for these properties was observed for supersaturated solutions.

Data for physical and transport properties of supersaturated solutions often are obtained by extrapolation of available data for saturated and undersaturated solutions into the supersatured region. This is often done empirically, since most available theory deals with dilute solutions of non-electrolytes (Reid, et al., 1977). Such extrapolation likely is appropriate where supersaturation levels are very low, as in most industrial crystallizations. For some circumstances, however, the supersaturation levels employed are very high (e.g., Narayanan, et al. (1982); Jagannathan, et al. (1980)) and the question arises whether extrapolation gives reliable property estimates. Data which address this question are meager.

Myerson, et al. (1982, 1984) recently measured the diffusivities for glycine, urea, KCl and NaCl in supersaturated and undersaturated solutions. The results in each case indicate that the diffusivity decreases precipitously once supersaturation is achieved. This effect has been associated with the influence of solute clusters (Cussler, 1980), the formation of which leads to homogeneous nucleation at sufficiently high supersaturations. Important evidence for the existence of such clusters has evolved. Larson and Garside (1986A) recently followed up work by Mullin and Leci (1969) to interpret concentration gradients in an isothermal column of supersaturated solution as due to solute clustering. They used an equilibrium thermodynamic model to estimate the number and size of solute clusters. They also (1986B) have argued that the stability of solute clusters of less than critical size is associated with size dependent interfacial energy. In addition, Raman spectroscopy studies by Ceretta and Berglund (1984) and Hussman et al. (1984) have provided evidence for the existence of clusters in supersaturated aqueous solutions of potassium nitrate, sodium nitrate, and ammonium dihydrogen phosphate.

These observations provide reasonable confirmation for the presence of clusters in supersaturated solutions. It seems likely that a physical property such as density will not be significantly affected by solute clustering, provided the solute molar volume in a cluster is not substantially different from unclustered solute and the cluster population is not extremely large. On the other hand, substantial effects might be expected for transport properties such as viscosity, electrical conductivity, and diffusivity. For dilute solutions, the Stokes-Einstein and Nernst-Einstein equations relate these properties. However, for concentrated solutions no adequate theory currently exists even at undersaturation. Thus the burden is still on experiment to provide key information.

H. Narayanan is now with the Polaroid Corporation, Waltham, MA.

In the present paper, some experimental results for the density, viscosity, and electrical conductivity for undersaturated and supersaturated aqueous solutions of several inorganic salts are presented. Five salts were chosen for study:
- sodium nitrate, which has high solubility, a relatively small temperature coefficient for solubility, forms a non-hydrated crystal, nucleates readily when supersaturated, and shows extreme dendritic crystal growth even at low supersaturations;
- potassium nitrate, which has a high solubility, a relatively large temperature coefficient for solubility, forms a non-hydrated crystal, nucleates readily when supersaturated, and shows extreme dendritic crystal growth even at low supersaturations;
- magnesium sulfate, which has a high solubility, a moderate temperature coefficient for solubility, forms a hydrated crystal, is stable toward nucleation at high supersaturations, and shows dendritic crystal growth at moderate supersaturations;
- potassium aluminum sulfate, which has a low solubility, a high temperature coefficient for solubility, forms a hydrated crystal, is very stable toward nucleation at high supersaturations, and does not exhibit dendritic growth of crystals;
- ammonium dihydrogen phosphate, which has a high solubility, a moderate temperature coefficient for solubility, forms a non-hydrated crystal, and is relatively stable toward nucleation at high supersaturations.

EXPERIMENTAL

Solution Preparation

For each system studied, stock solution was prepared by saturating at constant temperature in the presence of excess salt. After settling, the solution was decanted and stored at a temperature higher than saturation. Solution concentrations were measured by evaporation to constant weight. Less concentrated solutions were prepared by aliquot dilution.

Density Measurement

Solution density was determined by measuring the weight in solution of a glass plummet. A sample of solution was placed in a test tube immersed in a thermostated bath. The plummet was suspended from a Mettler balance mounted above the bath for ease in continuous weight measurement. Experiments were conducted at, above and below saturation by sequentially changing the temperature for a solution sample of constant composition.

Viscosity Measurement

Solution viscosity was measured using an Ostwald capillary viscometer immersed in a constant temperature bath. The viscometer was calibrated using distilled, deionized water. As with the density measurements, experiments were conducted sequentially by changing the temperature for a solution sample of constant composition.

Electrical Conductivity Measurement

Electrical conductivity was measured using an A.C. bridge circuit and a commercial conductivity cell. The cell was calibrated using 0.1 M. KCl and had a cell constant of about 1 cm^{-1}. The cell was fixed in a test tube containing solution and placed in a thermostated bath. As with the viscosity and density measurements, the experiments were conducted by sequentially changing the temperature for a solution sample of constant composition.

RESULTS

Density

Figure 1[*] shows the variation of specific volume for a potassium alum solution saturated at 22°C. (0.230 M.) over the temperature span 44 to 5°C. such that the relative saturation (C/C_s) ranges from 0.37 to 1.64. The specific volume changes monotonically with temperature (from supersaturation to undersaturation) and shows no evidence of any aberrations due to supersaturation. This is not surprising due to the rather low concentration of the salt. Figure 2 shows the molar volume for KNO_3 solutions versus solution composition at 25°C. Again, no unusual behavior in the supersaturated region is evident in spite of very high supersaturations. The results are similar to those of Urazov, et al. (1956) who concluded that density gives no clues to the differentiation between undersaturated and supersaturated solutions of sodium sulfate.

---

[*] Solutions referred to in the Figures are identified in Table 1.

From the properties of partial molar quantities,

$$\hat{V}_{KNO_3} = \hat{V}_s + (1-X_{KNO_3}) \frac{d\hat{V}_s}{dX_{KNO_3}}$$

The linearity of the plot suggests that the partial molar volume for KNO_3 is independent of concentration (over the range examined) and implies ideal solution behavior in the dilute solution sense. From the plot, $\hat{V}_{KNO_3} = 42.0$ cm^3/mole of KNO_3 in solution. For comparison, $\hat{V}_{KNO_3} = 47.4$ cm^3/mole when computed from the density of solid KNO_3.

Viscosity

Figure 3 shows the temperature dependence of viscosity for several solutions in the form suggested by the Andrade equation:

$$\mu = A \exp(B/T)$$

The deviation from linearity is rather small, and except for the magnesium sulfate solutions, the temperature dependence is very similar to that of water. Figure 4 consists of reference substance plots showing solution viscosity versus water viscosity at the same temperature. For potassium alum and ADP, the solution viscosity is linear with water viscosity, even at temperatures which correspond to very high supersaturations. For magnesium sulfate, by contrast, a significant deviation is noted in passing from the undersaturated to the supersaturated region. Moreover, the deviation from linearity increases significantly with increasing concentration. The reasons for this are unknown, but the implication is that a significant change in the level of solute clustering occurs when the solution becomes supersaturated and increases the solution viscosity accordingly. The effect is strong for magnesium sulfate which has a much higher solubility than either alum or ADP.

Electrical Conductivity

If significant enhancement of clustering occurs when a solution becomes supersaturated, one might expect this to be reflected in electrical conductivity as the concentration of ions is affected. Figure 5 shows the effect of temperature on the molar conductivity for the five systems studied, covering a range from undersaturation to very high supersaturation. For alum, KNO_3 and $NaNO_3$ the conductivity is essentially linear with temperature in the transition from undersaturation to supersaturation. For ADP and magnesium sulfate, however, a deviation appears.

The theory of electrical conductivity is moderately well developed (Smedley, 1980) and for dilute solutions the conductivity can be predicted exactly from the Onsager limiting law:

$$\Lambda_M = \Lambda^\circ_M + a\, C_M^{1/2}$$

For concentrated solutions the theory is shakier, but suggests that

$$\Lambda_M = \Lambda^\circ_M + f(C_M)$$

where $f(C_M)$ is a complex function of concentration which accounts for ion interactions, electrophoresis, incomplete dissociation and other effects. The temperature dependence for electrical conductivity arises principally through solvent viscosity (and to a lesser extent, the dielectric constant). For dilute solutions, Walden's Rule is at least approximately valid:

$$\Lambda_M \mu = \text{constant}$$

For concentrated solutions, the situation is more complicated but examination of the theory suggests that to a good approximation

$$1/\Lambda_M = A_1 + B_1 \mu$$

where A_1 and B_1 are concentration dependent constants. This implies that it should be possible to correlate the effect of temperature on conductivity by using the viscosity. It is not clear from theoretical arguments whether solvent viscosity or solution viscosity is more appropriate.

Figure 6 for potassium alum shows that, for this system, the reciprocal of the conductivity for a solution of constant composition is indeed linear with water viscosity, with no deviations resulting from the transition from undersaturation to very high supersaturation. Similar behavior for ADP and $NaNO_3$ was obtained. However, for magnesium sulfate, Figure 7, the plots are distinctly curved except for low concentrations. Since viscosity data were available for some of the solutions, the data were replotted in terms of solution viscosity. As Figure 8 shows, the data collapse nicely to a single curve. The deviation from linearity is slight except at high temperatures (i.e., low viscosities) where the effect of temperature on the dielectric constant may become important. When the data for alum was replotted in terms of solution viscosity, the result was similar except that significantly more scatter resulted.

For potassium nitrate, plots of the reciprocal of the conductivity versus water viscosity show distinct curvature as indicated by Figure 9. The trend is opposite to that shown by magnesium sulfate, showing a negative deviation from linearity, but with no distinct change in the transition to supersaturation. Solution viscosity data available for KNO_3 (and also ADP and $NaNO_3$) were insufficient to permit replotting of the data.

CONCLUSIONS

Data obtained for density, viscosity, and electrical conductivity for aqueous solutions of several inorganic salts show little evidence of unusual behavior in the transition to supersaturation. For viscosity and electrical conductivity this was a bit surprising in view of the extreme undercoolings examined (as much as 40°C, in some cases) and the large decrease in diffusivity demonstrated elsewhere for similar systems. No diffusivity data for supersaturated solutions of the types studied here are available currently, so the relations between viscosity, conductivity and diffusivity await further investigation. Water viscosity has been shown to be a useful correlating tool for the effects of temperature on both solution viscosity and electrical conductivity.

LITERATURE CITED

Cerretta, M. K., and K. A. Berglund, in Industrial Crystallization '84, S. J. Jancic and E. J. DeJong (Eds.), 233, Elsevier (1984).

Cussler, E. L., AIChE J 26, 43 (1980).

Hussman, G. A., M. A. Larson, and K. A. Berglund, in Industrial Crystallization '84, S. J. Jancic and E. J. DeJong (Eds.), 21, Elsevier (1984).

Jagannathan, R., C. Y. Sung, G. R. Youngquist, and J. Estrin, AIChE Symp. Ser. 193, No. 6, 90 (1980).

Larson, M. A., and J. Garside, Chem. Eng. Sci. 41, 1285 (1986).

Larson, M. A., and J. Garside, J. Crystal Growth 76, 88 (1986).

Mullin, J., and C. Leci, Phil. Mag. 19, 1075 (1969)

Myerson, A. S., and L. S. Sorell, AIChE J 28, 778 (1982).

Myerson, A. S., and Y. C. Chang, in Industrial Crystallization '84, S. J. Jancic and E. J. DeJong (Eds.), 27, Elsevier (1984).

Narayanan, H., G. R. Youngquist, and J. Estrin, J. Col. Int. Sci. 85, 319 (1982).

Reid, R. C., J. M. Prausnitz, and J. K. Sherwood, The Properties of Gases and Liquids, McGraw-Hill (1977).

Smedley, S. I., Interpretation of Ionic Conductivity in Liquids, Plenum Press, NY (1980).

Urazov, G. G., and L. S. Efimenko, J. Inorg. Chem. USSR 1, 100 (1956).

TABLE 1 - IDENTIFICATION OF SOLUTIONS

System	Code	Molar Conc.	Sat. Temp. °C
Alum	KA1	5.69×10^{-4}	
	KA2	2.04×10^{-3}	
	KA3	4.23×10^{-3}	
	KA4	1.48×10^{-2}	
	KA5	2.84×10^{-2}	
	KA6	5.30×10^{-2}	
	KA7	1.09×10^{-1}	
	KA8	1.52×10^{-1}	9.3
	KA9	1.66×10^{-1}	11.5
	KA10	2.14×10^{-1}	18.1
	KA10.1	2.27×10^{-1}	20.0
	KA11	2.77×10^{-1}	25.0
	KA12	3.84×10^{-1}	35.0
KNO_3	KN1	4.37×10^{-3}	
	KN2	4.37×10^{-2}	
	KN3	1.95×10^{-1}	
	KN4	4.37×10^{-1}	
	KN5	1.09	2.0
	KN6	2.19	13.5
	KN7	3.06	22.5
	KN8	4.37	36.0

TABLE 1 – Continued

System	Code	Molar Conc.	Sat. Temp. °C
$MgSO_4 \cdot 7H_2O$	MS1	1.045	
	MS2	1.578	
	MS3	1.798	0.0
	MS4	2.089	6.2
	MS5	2.520	16.3
	MS6	2.696	20.0
	MS7	2.925	26.2
	MS8	3.155	32.3
	MS9	3.334	37.1
	MS10	3.595	43.2
	MS11	3.867	48.8
ADP	AP1	2.96	24.4
	AP2	3.09	30.0
	AP3	3.54	35.6
$NaNO_3$	SN1	7.60	25.0

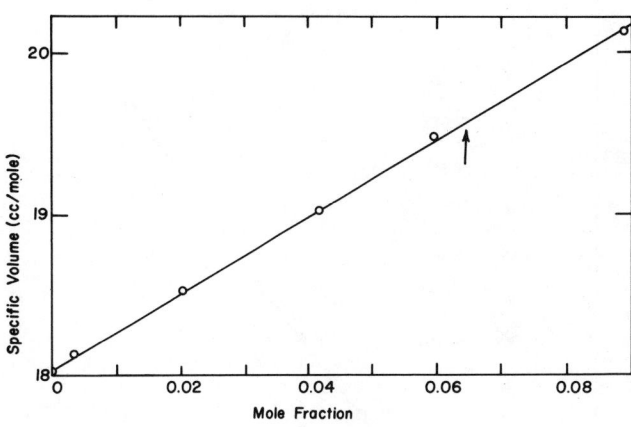

Figure 2. Specific volume vs mole fraction KNO_3 at 25°C. (arrow indicates saturation).

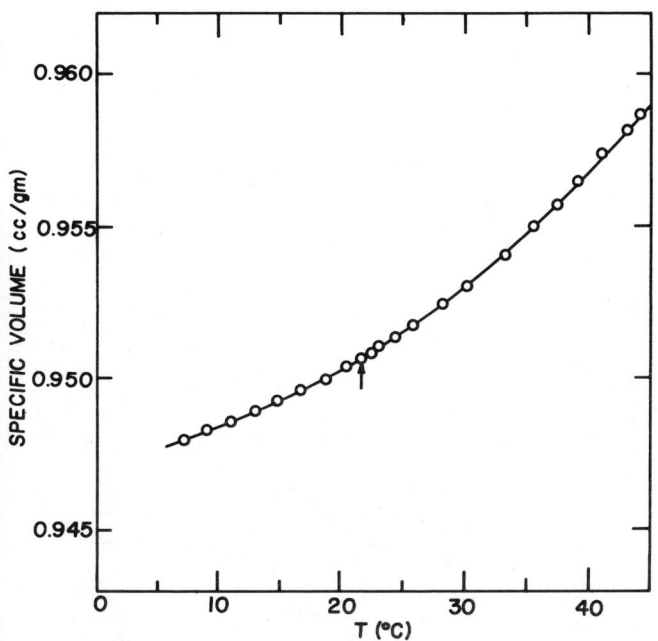

Figure 1. Specific volume vs temperature for potassium alum solution saturated at 22°C (0.230 M.) (arrow indicates saturation).

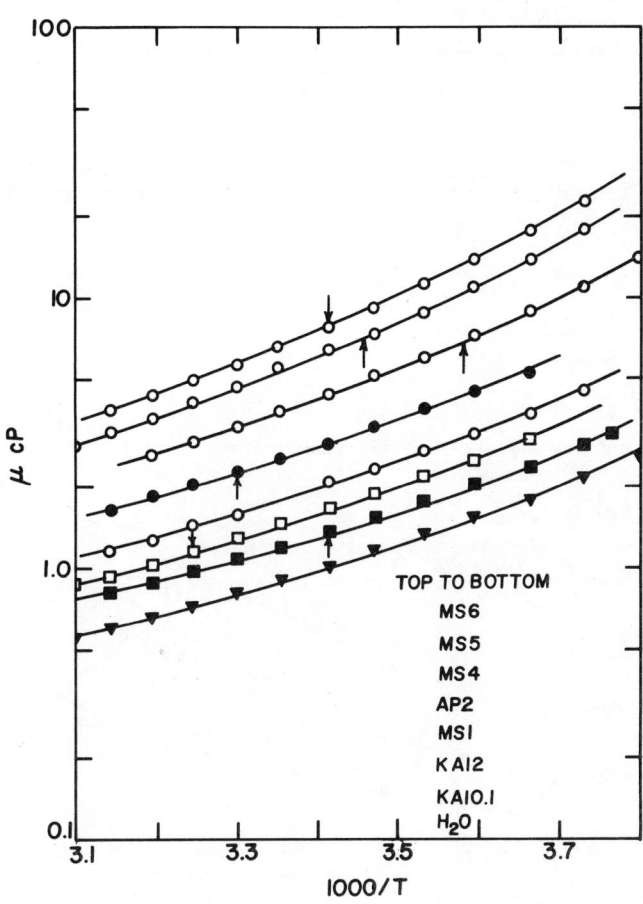

Figure 3. Temperature dependence of solution viscosity (arrows indicate saturation).

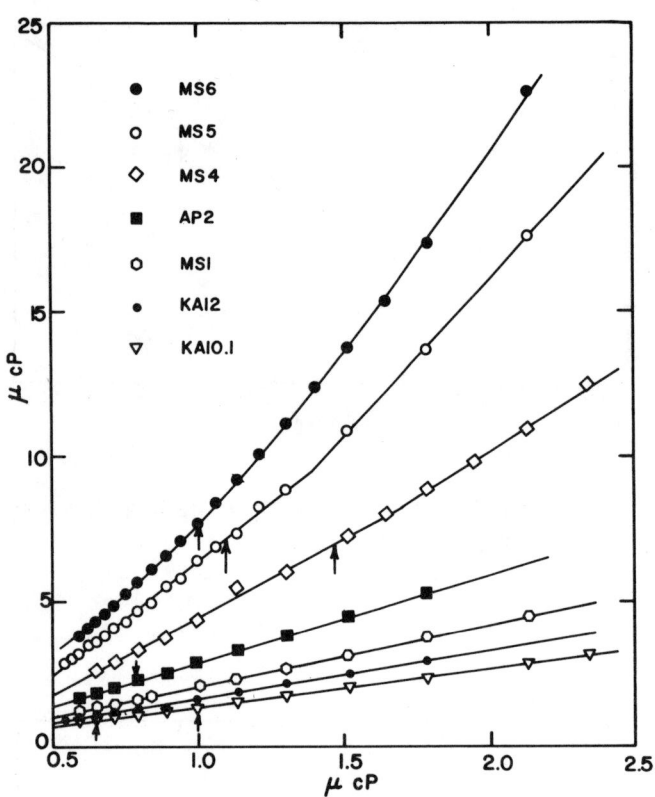

Figure 4. Solution viscosity versus water viscosity (arrows indicate saturation).

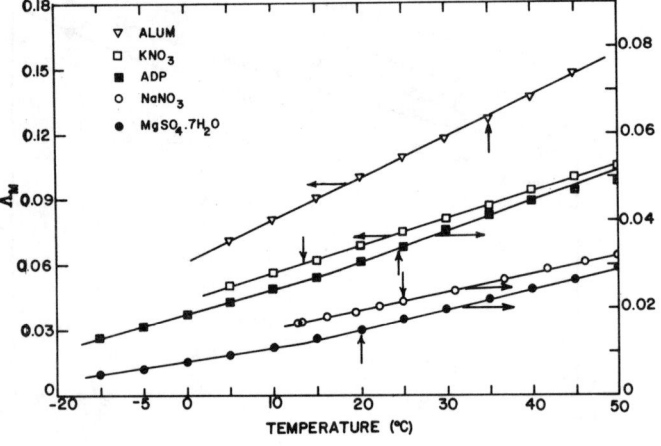

Figure 5. Temperature dependence of molar conductivity (vertical arrows indicate saturation).

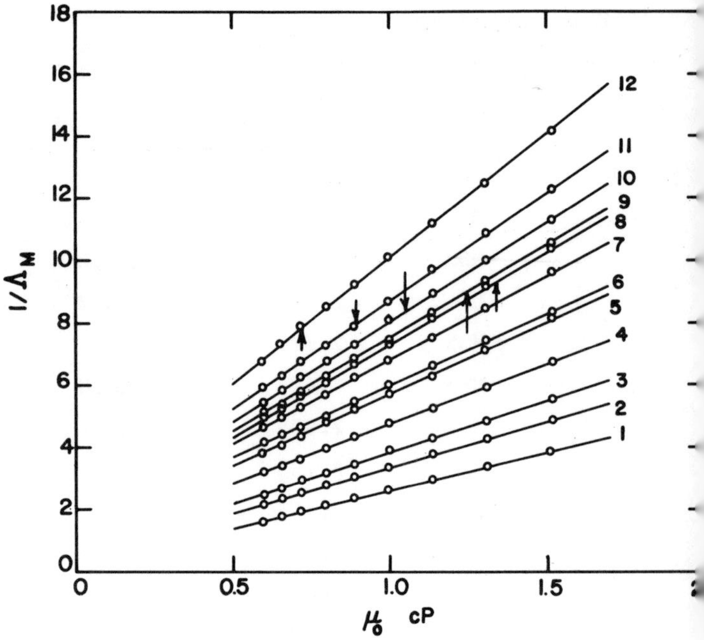

Figure 6. Reciprocal conductivity vs solvent viscosity for alum solutions (arrows indicate saturation).

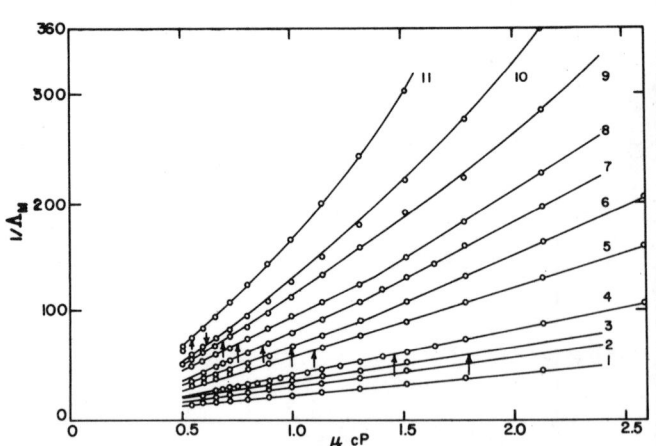

Figure 7. Reciprocal conductivity vs solvent viscosity for magnesium sulfate solutions (arrows indicate saturation).

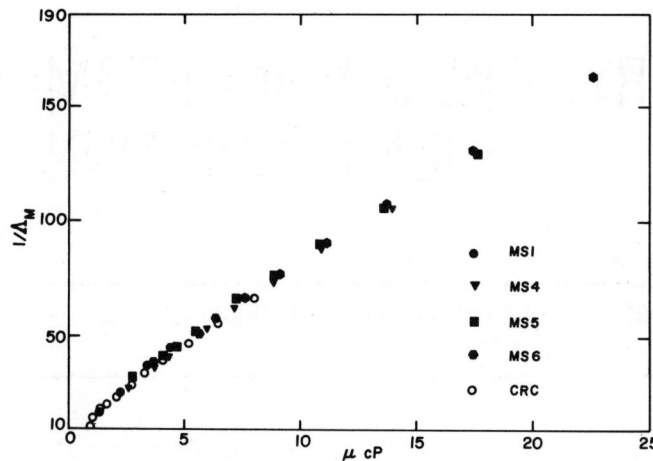

Figure 8. Reciprocal conductivity vs solution viscosity for magnesium sulfate.

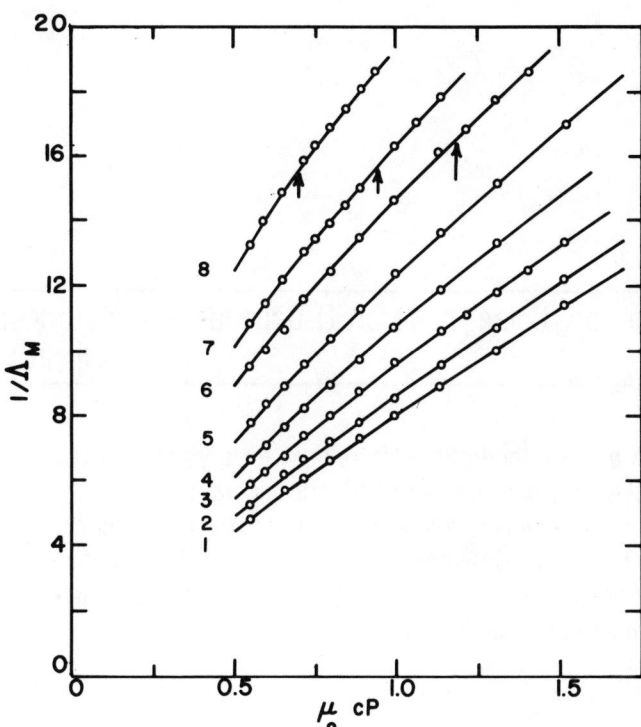

Figure 9. Reciprocal conductivity vs solvent viscosity for KNO_3 solutions (arrows indicate saturation).

CRYSTALLIZATION FROM IONIC SOLUTION

Pen-Pong Chiang and Marc D. Donohue ■ Department of Chemical Engineering, The Johns Hopkins University, Baltimore, MD 21218

A general theoretical framework for crystal growth is presented. Some important features of this framework are: (1) unlike previous theories, different adsorption isotherms are considered, including the isotherm obtained from electrical double layer theory; (2) several surface reaction mechanisms can be compared with experimental results; (3) new equations are formulated that not only fit experimental results but also provide a logical explanation for crystallization phenomena; and (4) a dimensionless ratio is presented which characterizes the transition from surface-reaction-controlled growth to diffusion-controlled growth.

The precipitation of ionic salts from solution is fundamentally different from most other phase transition processes because these crystals grow primarily by the addition of ions rather than molecules. For example, in the growth of silver chloride crystals from aqueous solution, the crystals normally grow by the addition of silver and chloride ions and not by the addition of silver chloride molecules that previously existed in solution. In this paper, we attempt to correlate data on crystal growth through kinetic mechanisms. We first discuss the growth of non-charged crystals, and then the growth of charged crystals.

Crystal growth from ionic solution occurs by a series of consecutive steps: bulk diffusion, surface adsorption, surface diffusion, surface reaction, and finally integration of ions or molecules into the crystal lattice. The first step, bulk diffusion, is a process which transports the individual ions from the bulk supersaturated solution to the crystal surface. The process depends linearly on the concentration gradient and inversely on diffusion layer thickness or crystal size (1-3).

The second step, surface adsorption, has not been widely investigated in the past. It has been reported that the adsorption of ions follows the forms of the Langmuir (4-5); Temkin (6), and Freundlich (7) isotherms. It also has been reported that one ion is adsorbed more strongly than the other on the crystal surface and, hence, the crystal retains a surface electrical charge and an electrical double layer is formed (8-9). Although there has been some discussion of the effect of adsorption on growth rate (10-11), there has been no systematic comparison of the rate expressions resulting from these different adsorption isotherms. Furthermore, there has been no discussion of the effect of preferential adsorption on crystal growth. In this paper, we derive growth rate expressions for ion adsorption which follows the Langmuir, Temkin, and Freundlich isotherms as well as the adsorption isotherm obtained from electrical double layer theory.

The third step, surface diffusion, has been discussed extensively in screw dislocation theories in relation to the growth of molecular crystals by vapor deposition (12-14). In these theories, the distance a molecule must diffuse in the adsorbed layer to a kink or growth site is shown to depend on both crystal size and supersaturation. Rather than consider this step separately, we ignore it in this paper. This is done because if surface diffusion were important, the rate expression resulting from the theory (which indicates a second-order relationship at low supersaturations and first-order relationship at high supersaturations) would be valid. This is not observed for most sparingly soluble salts where the growth rate data give a growth order of 2 or more even at high supersaturations (15-16).

The fourth step, surface reaction and the last step, integration of molecules or ions into the crystal lattice, can be either separate or interrelated, depending on the surface reaction mechanism. When oppositely charged ions are sequentially incorporated into the crystal lattice, the surface reaction and integration steps are combined, both physically and mathematically. Similarly, if a molecule becomes incorporated into the crystal lattice as soon as the ions react on the surface, these steps are combined also. However, when the molecules are formed on the crystal surface and then must diffuse to a kink or growth site to be incorporated into the crystal lattice, the two steps must be considered separately. While most theories of crystal growth have ignored the surface reaction and integration steps (12-14, 17-18), there have been attempts to explain experimental growth data by considering surface reaction/integration mechanisms (15). In this paper, we discuss several possible surface reaction and integration mechanisms in detail. We also show how to derive a general rate expression by combining these different surface reaction mechanisms with bulk diffusion and surface adsorption.

CRYSTAL GROWTH MECHANISMS

Crystal growth can occur by a variety of mechanisms. For example, it is obvious that crystals can grow by the addition of molecules or by the addition of individual ions that subsequently react and/or become incorporated into the crystal lattice. For systems that can grow by more than one mechanism, it is the fastest mechanism which determines the growth rate, but it is the slowest step in that fastest mechanism which controls the growth rate. If two or more steps in a mechanism have about the same rate, then the expression for the growth rate must reflect this. If two or more mechanisms have about the same overall rate, then the growth rate expression must reflect this also. Some possible crystal growth mechanisms are as follows:

Mechanism I: hereafter referred to as the Surface-Reaction/Molecule-Integration Mechanism (SR/MI)

$A^+(sol.) \longleftrightarrow A^+(ads.)$
$B^-(sol.) \longleftrightarrow B^-(ads.)$
$A^+(ads.)+B^-(ads.) \longleftrightarrow AB(ads.)$
$AB(ads.) \longleftrightarrow AB(lattice)$

Mechanism II: hereafter referred to as the Sequential-Ionic-Integration Mechanism (SII)

$A^+(sol.) \longleftrightarrow A^+(ads.) \longleftrightarrow A^+(lattice)$
$B^-(sol.) \longleftrightarrow B^-(ads.) \longleftrightarrow B^-(lattice)$

where A^+ and B^- stand for the positive and negative ions, respectively. Other possible growth mechanisms occur when a crystal grows by the addition of molecules that were formed previously in solution, i.e.,

Mechanism IIIA: Molecular Growth

$A^+(sol.)+B^-(sol.) \longleftrightarrow AB(sol.)$
$AB(sol.) \longleftrightarrow AB(lattice)$

Mechanism IIIB: Molecular Growth

$A^+(sol.)+B^-(sol.) \longleftrightarrow AB(sol.)$
$AB(sol.) \longleftrightarrow AB(ads.) \longleftrightarrow AB(lattice)$

Since the concentration of AB molecules in solution usually is much lower than the concentrations of the ions, this is not a common growth mechanism. This mechanism is likely only when AB is the dominant species (e.g., in silver chloride solution with a slight excess of chloride ion). In the following derivation, we assume that growth by mechanisms IIIA and IIIB is negligible compared to growth by the addition of A^+ and B^- ions.

THEORY OF CRYSTAL GROWTH

Without Considering Surface Adsorption

The expression for the net flux of individual ions by diffusion to the crystal solution interface is given as (1-3),

$$G = \frac{dr}{dt} = \frac{D}{\rho}\frac{dC}{dx} = D\frac{(C_o-C_i)}{\rho r} \qquad (1)$$

where D = diffusion coefficient, ρ = crystal density, r = crystal radius, C_o = bulk concentration, and C_i = interfacial concentration.

The mass transfer coefficient is, in general, defined as M=D/ρr, and Equation (1) can be simplified to, G=M(C_o-C_i). Following Equation (1), the net flux based on the diffusion process for species A^+ and B^- are

$$G_a = M_a([A^+]_o-[A^+]_i) \qquad (2)$$

$$G_b = M_b([B^-]_o-[B^-]_i) \qquad (3)$$

where $[A^+]_o$ and $[B^-]_o$ are the bulk concentrations for species A and B, and $[A^+]_i$ and $[B^-]_i$ are the interfacial concentrations for each species.

Since the diffusion process should maintain stoichiometric proportions in order to maintain the condition of electroneutrality, the two net fluxes must be equal for the case of 1,1 electrolytes. That is,

$$G_{ab} = G_a = G_b \qquad (4)$$

The rate of incorporation of individual adsorbed ions into the crystal lattice by surface reaction is determined by the surface reaction mechanism. Mechanisms I and II, described above, are treated separately in order to show their differences.

Mechanism I: Surface-Reaction/Molecule-Integration

For this mechanism, Chiang and Donohue (19) derived the growth rate in a supersaturated solution as,

$$G = f - b = K_I([A^+]_i [B^-]_i - [A^+]_e [B^-]_e) \qquad (5)$$

where K_I is a surface reaction/integration constant, $[A^+]_e$ and $[B^-]_e$ are the equilibrium concentrations for each species.

At steady state, since the flux by diffusion must equal that by surface reaction, Equations (2), (3), and (5) can be combined, thus eliminating $[A^+]_i$ and $[B^-]_i$. The result is,

$$\frac{K_I G^2}{M_a M_b} - (1 + \frac{K_I[A^+]_o}{M_a} + \frac{K_I[B^-]_o}{M_b})G \qquad (6)$$

$$+ K_I([A^+]_o [B^-]_o - [A^+]_e [B^-]_e) = 0$$

Equation (6) is a general form which allows one to calculate the rate of growth not only for equivalent concentrations but for non-equivalent concentrations. Consider a simple case where there are equivalent concentrations and where the mass transfer coefficients for both species are the same, i.e., $[A^+]_o = [B^-]_o$ and $M_a = M_b$. Then Equation (6) reduces to,

$$K_I \frac{G^2}{M_a^2} - (1 + \frac{2K_I[A^+]_o}{M_a})G + K_I([A^+]_o^2 - [A^+]_e^2) = 0 \qquad (7)$$

In the case $2K_I[A^+]_o/M_a \gg 1$, Equation (7) yields,

$$G = M_a([A^+]_o - [A^+]_e) \qquad (8)$$

which is diffusion-controlled growth. On the other hand, if $K_I[A^+]_o/M_a \ll 1$, and $K_I \ll M_a$, the growth is controlled by surface reaction (strictly speaking, it is surface reaction/integration), and is expressed as,

$$G = K_I([A^+]_o^2 - [A^+]_e^2) \qquad (9)$$

For those conditions which are between these two limiting cases, Equation (6) must be solved for G. The results can be found in Chiang (20).

Mechanism II: Sequential-Ionic-Integration

A general expression for the growth rate of non-charged crystals, according to Mechanism II, is obtained by Chiang and Donohue (19).

$$(\frac{[N]K_{A,II}K_{B,II}}{M_a M_b} + \frac{K_{A,II}}{M_a} + \frac{K_{B,II}}{M_b})G^2 - \qquad (10)$$

$$(\frac{[N]K_{A,II}K_{B,II}[B^-]_o}{M_a} + \frac{[N]K_{A,II}K_{B,II}[A^+]_o}{M_b}$$

$$+ K_{A,II}[A^+]_o + K_{A,II}[A^+]_e + K_{B,II}[B^-]_o + K_{B,II}[B^-]_e)G$$

$$+ [N]K_{A,II}K_{B,II}([A^+]_o [B^-]_o - [A^+]_e [B^-]_e) = 0$$

where $K_{A,II}$ and $K_{B,II}$ are the forward rate constants for cations and anions, [N] is the surface concentration or number density of lattice growth sites. Equation (10) can be applied to both equivalent and non-equivalent ionic concentrations. Again, consider the simple case of letting $M_a = M_b$, $[A^+]_o = [B^-]_o$, $[A^+]_e = [B^-]_e$, and $K_{A,II} = K_{B,II}$. Then Equation (10) reduces to,

$$(2 + \frac{K_{A,II}[N]}{M_a})G = K_{A,II}[N]([A^+]_o - [A^+]_e) \qquad (11)$$

Again, there are two limiting cases in Equation (11): for $K_{A,II}[N]/M_a \ll 2$

$$G = \frac{K_{A,II}[N]([A^+]_o - [A^+]_e)}{2} \qquad (12)$$

and for $K_{A,II}[N]/M_a \gg 2$

$$G = M_a([A^+]_o - [A^+]_e) \qquad (13)$$

The first case corresponds to surface-reaction-controlled growth while the second case is for diffusion-controlled growth. The interesting result is that the growth is linear in both limits.

Langmuir, Temkin, and Freundlich Adsorption Isotherms

The above derivation dealt with an ideal case where no adsorption was considered. In practice, the surface reaction is determined by the adsorbed surface concentration, and hence the adsorption isotherm should be considered. It has been shown by Weisz (21) that at steady state the interfacial concentration, which is determined by both the diffusion and surface adsorption processes, is often determined by the diffusion process. Further, since the adsorbed concentration is a function of the interfacial concentration, it also is controlled by the diffusion process. Therefore, Equations (2) and (3) are still applicable when adsorption is considered.

Chiang and Donohue (19) took different surface adsorption isotherms into account in deriving growth rate expressions in a supersaturated solutions. Using

Langmuir isotherm as an example, the growth rate according to Mechanism I is given by,

$$G = \frac{K_I'[A^+]_i[B^-]_i}{(1+K_{AL}[A^+]_i+K_{BL}[B^-]_i)^2} - \frac{K_I'[A^+]_e[B^-]_e}{(1+K_{AL}[A^+]_e+K_{BL}[B^-]_e)^2} \quad (14)$$

where K_{AL} and K_{BL} are adsorption constants for positive and negative ions based on Langmuir adsorption isotherm.

For most sparingly soluble salts, the term $(K_{AL}[A^+] + K_{BL}[B^-])$ in the denominator can be eliminated since the concentration of each species is very low (in the range of 10^{-4} to 10^{-7} M). The expression for the growth rate can be simplified to,

$$G = K_I'([A^+]_i[B^-]_i - [A^+]_e[B^-]_e) \quad (15)$$

which is exactly the same as that derived without having considered the adsorption isotherm (Equation 5).

For other types of adsorption isotherms and mechanisms, one can derive similar rate expressions. Since the derivation is essentially the same, we will not repeat the procedure. The results can be found in Chiang (20).

Growth Rate Expression from Electrical Double Layer Theory

When the electrical double layer is considered for a charged crystal, not only the diffusion process but also the electromigration process should be taken into account. The counter-ions, say positive ions, should migrate toward the surface under the action of the electric field, but because their concentration increases in this direction, this electromigration flux is counterbalanced by the diffusion flux. For the negative ions, the diffusion process is toward the surface, but this is opposed by the electromigration process. The surface reaction/integration process is assumed to be controlled by the sequential-ionic-integration mechanism. Because the negative ions are adsorbed right on the surface, while the positive ions are located in an immobilized layer which is adjacent to the surface, the positive ions and negative ions are separately incorporated into the crystal lattice. One may assume the growth sites for the positive ions are the same as the adsorbed concentration of negative ions.

Because of the existence of the double layer, the backward rate is approximated from the equilibrium state: the concentration of counter-ion in the outer layer is in equilibrium with the bulk concentration, while the concentration of co-ion on the crystal surface remains the same. The backward rate in this case is, The growth rate is therefore (19),

$$G = K_{II}(\ln\frac{[B^-]_o}{[B^-]_{pzc}} - \frac{zF}{kT}\zeta)([A^+]_o - [A^+]_e) \quad (16)$$

where $[B^-]_{pzc}$ is the concentration for the negative ion at the point of zero charge (abbreviated as pzc), z is the valence of the ion, F is Faraday's constant, k is Boltzmann's constant, T is temperature and ζ is zeta potential.

RESULTS AND DISCUSSION

In order to compare the theory with the experimental findings, one first must define supersaturation. In this paper, we define the supersaturation ratio as the ratio of the bulk concentration to the saturated concentration. In a 1,1 electrolyte aqueous solution, $S = C_o/C_e = ([A^+]_o[B^-]_o/[A^+]_e[B^-]_e)^{1/2}$. In n-m electrolyte solutions, the supersaturation ratio is expressed as $S = ([A^+]_o^n[B^-]_o^m/[A^+]_e^n[B^-]_e^m)^{1/(n+m)}$.

Several surface reaction models (10-11, 15, 22-23) were introduced to explain experimentally observed growth rates of the form,

$$G = K(S-1)^p \quad (17)$$

where S is the supersaturation ratio, and p is the growth-rate order (see Table 1). Many experiments involving spontaneous crystallization have been conducted to measure the order of the growth rate. McCabe, among the first to measure the rate of crystal growth for copper sulfate, found p=1 (39). He suggested that diffusion was the controlling step. On the other hand, Davies and Jones (10) found second-order kinetics for the growth of silver chloride. Other experiments indicated that the over-all growth rate order is in the range of 1 to 2 (37-38, 45-46). A few experiments showed that the order could be greater than 2 (16,28). Several measured values of the growth rate order for different substances are given in Table 1. From Table 1 one may conclude that for most slightly soluble substances, the rate of growth is generally second-order, even at high supersaturations. For soluble salts, the growth rate usually follows the first-order dependence. Some measurements on soluble salts indicate that the growth rate changes from a second-order to a first-order as the supersaturation increases.

To show how well the crystal growth theory, derived from different types of adsorption isotherms and different surface reaction/integration mechanisms, predicts the experimental findings, we discuss each of them individually. First, we consider the expression for crystal growth obtained from the linear adsorption isotherm. When the sequential-ionic-integration mechanism is considered, the growth rate expression always gives a first-order dependence of the rate of growth on supersaturation, as shown in Figure 1a. Figure 1a is made by arbitrarily assuming a mass transfer coefficient of 10 cm/sec M^{-1} and varying the surface reaction rate constants from 10^6 to 10^{-3} cm/sec M^{-2}. When the surface reaction rate constant is much higher than the mass transfer coefficient, the growth is diffusion limited, as denoted by line (a) in Figure 1a. On the other hand, growth is surface reaction limited when the mass transfer coefficient is much higher than the surface reaction constant, as represented by line (d). Unfortunately, this first-order dependence is inconsistent with experimental findings, especially for sparingly soluble salts.

For the surface-reaction/molecule-integration mechanism, we obtain some interesting results concerning the order of its growth rate. To show this, we use the same sets of constants in Figure 1b. Diffusion control occurs when the mass transfer coefficient is much smaller than the surface reaction constant and the rate expression follows a first-order dependence of supersaturation, as represented by line (a) in the figure. When the surface reaction rate constant is comparable to or lower than the mass transfer coefficient, e.g., curves b, c, and d in the figure, transitions from first-order to second-order to first-order are observed; the growth rate follows a first-order dependence at extremely low supersaturations, a second-order dependence at medium to high supersaturations, and a first-order dependence again at extremely high supersaturations.

As was shown before, if the dimensionless ratio $2K_I[A^+]_o/M_a$ greatly exceeds 1.0, the expression for the growth rate follows the form of diffusion control. On the other hand, when $2K_I[A^+]_o/M_a$ is far less than 1.0, it will follow surface reaction control. The ratio K_I/M_a generally varies from 10^1 to 10^4, depending on the system (20). The magnitude of the dimensionless function $(K_I[A^+]_o/M_a)$ is determined by the concentration. The higher the concentration, the greater is the value $K_I[A^+]_o/M_a$. In soluble salts, the concentration is usually very high. This high concentration provides the explanation why, for most soluble salts, a first-order dependence of the growth rate on supersaturation generally is observed. (Another possible explanation for this first-order dependence is a result of sequential-ionic-integration mechanism, as discussed above.) Since the concentrations are usually very low for slightly soluble or sparingly soluble salts, the dimensionless ratio is far less than 1.0. For these systems, the growth process is controlled by surface reaction.

The above discussion is made for a rate expression which was derived for the linear adsorption isotherm. When another type of adsorption isotherm is considered, the general form of growth rate can be calculated in terms of a given set of constants for the mass transfer coefficient and the surface reaction rate constant. Using the Langmuir adsorption isotherm and assuming the ions are adsorbed in equivalent amounts, we obtain the numeric solution of growth rate with given constants. We find that the results, calculated from both surface-reaction/molecule-integration mechanism and sequential-ionic-integration mechanism considering the Langmuir isotherm, are essentially identical to those we calculated for the linear adsorption isotherm. Therefore, the above conclusions, made for the linear adsorption isotherm, also should be true for the case when the Langmuir isotherm is used. In the case of non-equivalent adsorption, we find that, when $K_{AL}=0.1K_{BL}$, there is no significant change in the growth-rate order.

For other types of adsorption isotherms, e.g., for the Freundlich isotherm, the rate of growth depends on the constant n. Specifically, the growth rate for surface reaction control due to surface-reaction/molecule-integration mechanism is,

$$G=K([A^+]_o^{2/n}-[A^+]_e^{2/n}) \qquad (18)$$

If n equals 1, Equation (18) is identical to that derived from either considering the Langmuir isotherm or without considering any type of adsorption isotherm. In the case when n is equal or greater than 2, order of the growth rate is less than unity. Since n in most cases is greater than 1 (4), we have used a value of 1.5 in our calculations. By using this value, we find that the growth rate order is 1.2 for surface reaction molecule/integration mechanism (20). This is also inconsistent with the experimental findings and therefore the isotherm will not be discussed further. The same conclusion can be drawn for sequential-ionic-integration mechanism.

To compare the derived rate expression, $G=K([A^+]_o^2-[A^+]_e^2)$, with the experimental results, we use silver chloride, calcium carbonate, barium sulphate, and lead chromate as examples. These experiments, conducted at conditions from extremely low supersaturations (silver chloride) to extremely high supersaturations (lead chromate), all were reported to have a second-order dependence of growth rate on supersaturation. Since the experiment was conducted at very low supersaturation, the derived rate expression follows a first-order relationship (as shown in Figure 1b) and thus is far from the growth measurements. Figure 2 shows the comparison of the growth rate expression with the results for calcium carbonate (43). Since supersaturation in this experiment ranged from low to medium, the rate expression changes from first-order to second-order, as shown in Figure 2 (also in Figure 1b). When supersaturation is higher than two, the rate expression has good agreement with the second-order dependence. Figure 3 shows the comparison of the growth rate expression with the results for lead chromate taken by Packter (59). As evident from the figure, at such high supersaturations, the rate expression is identical to the second-order relationship. Another comparison was made by Marshall and Nancollas (60). They used the expression to interpret the crystal growth of $CaHPO_4 \cdot 2H_2O$, and an excellent agreement was obtained.

In the following, we compare the growth rate expression, derived from electrical double layer theory, with experimental data. Silver chloride and barium sulphate are used as two examples for comparison. We use Equation (16) for comparison.

In order to compare the rate expression with the experimental data, two important parameters must be known: the zeta potential and the point of zero charge. For silver chloride, the point of zero charge for the chloride ion occurs at $pCL^- = 5.17$ (61), which is equivalent to $[B^-]_{pzc} = 10^{-5.17} = 6.16 \times 10^{-6} M$. The zeta potential at a concentration of $1.55 \times 10^{-5} M$ is extrapolated to be -17.6 mv (62). When these values are used in Equation (26), there is good agreement between the present work and the experimental results of Davies and Jones (10), as shown in Figure 4. In the same figure, we also plot the growth rate using Equation (9), and this is far from the measurements.

Note that the zeta potential in this calculation can be considered a constant since the electrolyte concentration change in the experiment is small. When this is true, a simplification for Equation (16) can be obtained by substituting the exact values for $[B^-]_{pzc}$ and ζ. After this simplification, we find that the term $\ln \frac{[B^-]_o}{[B^-]_{pzc}} - (zF/kT)\zeta$ is approximately equal to $([B^-]_o - [B^-]_e)/[B^-]_e$, where $[B^-]_e = 1.328 \times 10^{-5}$ M. This simplification is equivalent to saying that the surface adsorbed concentration for the co-ion is proportional to $[B^-]_o - [B^-]_e$, and Equation (16) is therefore approximated by,

$$G \approx K_{II}([B^-]_o - [B^-]_e)([A^+]_o - [A^+]_e) \tag{19}$$

This result has been observed in experiments by several authors (2, 10-11). It was also postulated by Davies and Jones, however, they could not derive or justify this result. When the bulk concentration of each species is the same, Equation (19) reduces to a widely observed second-order dependence of the growth rate on supersaturation.

For barium sulphate, probably the most extensively studied sparingly soluble salt, several different results have been observed. Turnbull (31) observed third-order kinetics, while O'Rourke and Johnson (32) reported fourth-order growth. Gunn and Murthy (28) also reported a third-order dependence on supersaturation when the crystal was small; however, for large crystals at low supersaturation they observed the dependence was reduced to first-order. Nancollas and Purdie (35), Hostomsky et al. (29), and Walton and Hlabse (33), found second-order dependence for growth in seeded solution, while Collins and Leinweber (34) reported agreement of their experimental results with an equation based on first-order growth. These disparate results demonstrate the difficulty of finding a simple interpretation for growth of barium sulphate.

For barium sulphate, Equation (16) is inappropriate since the double layer is reversely charged, i.e., the co-ion is barium, and counter-ion is sulphate (63). In this case, we must rewrite Equation (16) as,

$$G = K'_{II}(\ln \frac{[A^+]_o}{[A^+]_{pzc}} - \frac{zF}{kT}\zeta)([B^-]_o - [B^-]_e) \tag{20}$$

The corresponding data for this calculation are, $K_{sp} = 9.86$ and $P_{pzc} = 3.1$ (64). We assume that the zeta potential is a constant, 64mv (65) for the supersaturation range from 5 to 50. Figure 5 shows the comparison of the growth rate with the results obtained by Hostomsky et al. (29), which is reported to have a second-order dependence on supersaturation. We find that the expression, obtained from the electrical double layer theory, agrees with the second-order dependence. Figure 5 also includes a rate expression which is proportional to $([A^+]_o^2 - [A^+]_e^2)$.

One difficulty in using the electrical double layer theory is uncertainty in the value of the zeta potential. In the above calculation the zeta potential was assumed to be a constant. This assumption is valid only when supersaturation does not change much. For the silver chloride experiments, the supersaturation was in a narrow range (1.09-1.28) and thus this assumption was reasonable. However, this is not the case for barium sulphate, when the supersaturation covered a much wider range of values (5-50), and, therefore, the zeta potential may not have been constant. Since we know that the adsorbed concentration for co-ions is proportional to the difference of the surface potential and the zeta potential, we had hoped to find experimental data which relates the surface potential and the zeta potential to the electrolytic concentration. Unfortunately, no such data are available. Overbeek (66) using silver iodide, calculated the change of both the zeta potential ζ and the surface potential, ψ_o, as a function of pAg. These curves are shown in Figure 6. The parallelism between the two curves near the pzc is apparent. Far from the pzc, the deviation increases with increasing concentration. In a certain range, ζ decreases after reaching a maximum. Therefore, if experiments are performed in this regime, the surface adsorbed concentration should relate to $([A^+]_o - [A^+]_e)$ by a higher-order, not first-order. Consequently, this would explain growth-rate orders higher than two. Further, as pointed out before, the point of zero charge is not an intrinsic constant, but depends on various factors, including the crystal size, pH, concentration of other dissolved salts and measurement technique. For example, for barium sulphate P_{pzc} is 3.1 using the streaming potential method (64), 3.9 using the electro-osmosis method (67), while it is 3.6 using the microelectrophoresis method (68). According to Barr and Dickinson (69), the pzc of a crystal of AgBr decreased as the crystal size increased.

The rate of growth is merely the difference between the forward and backward rates (G = f-b). The reverse process of crystal growth, dissolution, can be expressed using a similar approach. Dissolution rate measurements are made in such a manner that the seed crystals are added to an undersaturated solution. In this case, the dissolution rate is equal to the difference of the backward rate and forward rate, i.e., D = b-f.

Following the derivation for the growth rate, the rate of dissolution of crystals according to surface-reaction/molecule-integration mechanism can be written in the form,

$$\frac{K_I}{M_a^2}D^2 + (\frac{2K_I[A^+]_o}{M_a} - 1)D + K_I([A^+]_e^2 - [A^+]_o^2) = 0 \quad (21)$$

The dimensionless ratio $(2K_I[A^+]_o/M_a)$ plays similar role in the rate determining step. When this ratio is far less than 1.0, it is the surface reaction control and $D = K_I([A^+]_e^2 - [A^+]_o^2)$. The controlling step is diffusion if the ratio greatly exceeds 1.0 and the rate of dissolution becomes $D = M_a([A^+]_e - [A^+]_o)$. As was pointed out before, for soluble salts the function is usually far greater than 1.0. This explains the general observation that dissolution for soluble salts is more or less entirely diffusional in nature (70).

CONCLUSIONS

We presented a theory for the growth of non-charged crystals and discussed how it can be applied to two different growth mechanisms. The theory can be applied readily to other systems or mechanisms with only minor modifications. The growth rate expression, derived for non-charged crystals, explains several sets of experimental data very well. The expression also qualitatively explains the transition, i.e., parabolic growth at low to medium supersaturation and linear growth at high supersaturation. We cannot, however, use the expression to explain the results for silver chloride. This discrepancy indicates that at very low supersaturation, crystal charge or other effects must be involved.

We also derived a rate expression for charged crystals. With this expression, we can explain the widely observed second order dependence of growth rate on supersaturation. We also provide possible explanations for cases where the growth rate has a higher-order dependence on supersaturation.

ACKNOWLEDGEMENTS

We would like to express our gratitude to Dr. P. Vimalchand for his help in preparation of this paper.

LITERATURE CITED

1. Walton, A.G., The Formation and Properties of Precipitants, Interscience, New York (1967)

2. Mullin, J.W., Crystallization, Butterworths, London, (1972)

3. Nernst, W., Z. Phys. Chem. 47, 52 (1904)

4. Hayward, D.O., and B.M. Trapnell, Chemsorption, Butterworth Pub., (1964)

5. Laidler, K.J., and J.H. Meiser, Physical Chemistry, The Benjamin/Cummings Publishing Comp. Inc., (1982)

6. Brunauer, B., K.S. Love and R.G. Keenan, J. Am. Chem. Soc., 64, 751 (1942)

7. Freundlich, H.M.F., Kappilarchemie, Leipzig, (1909)

8. Grahame, D.C., Chem. Rev., 41, 441 (1947)

9. Verwey, E.J.W., and J.Th.G. Overbeek, Theory of the stability of leophobic colloid, elsevier, (1948)

10. Davies, C.W., and A.L. Jones, Trans. Faraday Soc., 51, 812 (1955)

11. Walton, A.G., J. Phys. Chem., 67, 1920 (1963)

12. Burton, W.K., N. Cabrera, and F.C. Frank, Phil. Trans., A243, 299 (1951)

13. Chernov, A.A., Soviet Phys. Usp., 4, 116 (1961)

14. Gilmer, G.H., R. Ghez, and N. Cabrera, J. Crystal Growth, 8, 79 (1971)

15. Doremus, R.H., J. Phys. Chem., 62, 1068 (1958)
16. Doremus, R.H., J. Phys. Chem., 74, 1405 (1970)
17. Becker, R., Ann. Phys., 32, 128 (1938)
18. Bradley, R.S., Quart. Rev. (London) 4, 315 (1951)
19. Chiang, P.P. and M.D. Donohue, A Kinetic Approach to Crystallization from Ionic Solution, submitted to J. Colloid & Int. Sci., (1986)
20. Chiang, P.P., A Kinetic Approach to Crystallization from Ionic Solution, PhD thesis, Johns Hopkins University (1985)
21. Weisz, P.B., Trans. Faraday Soc., 63, 1801 (1967)
22. Humphreys-Owen, S.P.F., Proc. R. Soc., A197, 218 (1949)
23. Nielsen, A.E., Pure & Appl. Chem., 53, 2025 (1981)
24. Klein, Haneveld H.B., J. Crystal Growth, 10, 111 (1961)
25. Botsaris, G.D., E.A. Mason, and R.C. Reid, J. Chem. Phys., 45, 1893 (1966)
26. Rumford, F., and J. Bain, Trans. Inst. Chem. Engrs., 38, 10 (1960)
27. Barnone, J.P., D. Svrjcek, and G.H. Nancollas, J. Crystal Growth, 62, 27 (1983)
28. Gunn, D.J., and M.S. Murthy, Chem. Eng. Sci., 27, 1293 (1972)
29. Hostomsky, J., J. Rathousky, and J. Skrivanek, J. Crystal Res. & Tech., 16, 759 (1981)
30. Nancollas, G.H., and N. Purdie, Trans. Farad. Soc., 59, 735 (1963)
31. Turnbull, D., Acta Metallurgical, 1, 684 (1953)
32. O'Rourke, J.D., and R.A. Johnson, Anal. Chem., 27, 1699 (1955)
33. Walton, A.G., and T. Hlabse, Talanta, 10. 601 (1963)
34. Collins, F.C., and J.P. Leinweber, J. Phys. Chem., 60, 389 (1956)
35. Nancollas, G.H., and N. Purdie, Quart. Rev., 18, 1 (1964)
36. Garside, J., R. Janssen-Van Rosmalen, and P. Bennema, J. Crystal Growth, 29, 353 (1966)
37. Garside, J., and J.W. Mullin, Trans. Inst. Chem. Engs., 46, T11 (1968)
38. McCabe, W.L., Ind. Eng. Chem., 21 30 (1928)
39. McCabe, W.L., and R.P. Stevens, Chem. Eng. Prog., 47, 168 (1951)
40. Nancollas, G.H., and N. Purdie, Trans. Farad. Soc., 57, 2272 (1961)
41. Campbell, J.R., and G.H. Nancollas, J. Phys. Chem., 73, 1735 (1969)
42. Nancollas, G.H., J. Crystal Growth, 42, 185 (1977)
43. Nielsen, A.E., and J.M. Toft, J. Crystal Growth, 67, 278 (1984)
44. Mullin, J.W., and A. Amatavivadhana, J. Appl. Chem., 17, 151 (1967)
45. Kahlweit, M., J. Crystal Growth, 3-4, 401 (1968)
46. Reich, R., and Kahlweit, M., Ber. Bunsenges Phys. Chem., (Ger.) 72, 70 (1968)
47. Khamskii, E.V., and Podozerskaya, E.A., J. Apply. Chem., (USSR) 42, 1829 (1969)
48. Timm, D.C., and T.R. Cooper, AICHE J., 17, 285 (1971)
49. Van Hook, A., J. Phys. Chem., 44, 751 (1940)
50. Howard, J.R., and G.H. Nancollas, Trans. Farad. Soc., 53, 1449 (1957)
51. Melikhov, I.V., E.K. Kirkova, and M.D. Djarova, J. Crystal Growth, 53, 547 (1981)
52. Liu, S.T., and G.H. Nancollas, J. Crystal Growth, 6, 281 (1970)
53. Christoffersen, M.R., J. Christoffersen, M.P.C. Weijnen, and G.M. Van Rosmalen, J. Crystal Growth, 58, 585 (1982)
54. Smith, B.R., and F. Sweett, J. of Colloid & Int. Sci., 37, 612 (1961)
55. Tadros, M.E., Skalny, J., and R.S. Kalyoncu, J. Colloid & Int. S., 55, 20 (1976)
56. Mullin, J.W., and C. Gaska, Can. J. of Chem. Eng., 47, 483 (1969)
57. DeSilva, R.L., and D.E. Creasy, Crystal Res. and Technol., 18, 6, 725 (1983)
58. Reddy, M.M., J. Crystal Growth, 41, 287 (1977)
59. Packter, A., J. Chem. Soc., (A) 859 (1968)
60. Marshall, R.W., and G.H. Nancollas, J. Phys. Chem., 73, 3838 (1969)
61. Hoyen, H.A., and R.M. Cole, J. Colloid & Int. Sci., 41, 93 (1972)
62. Insley, M.J., and G.D. Parfitt, Trans. Faraday Soc., 64, 1945 (1968)

63. Honic, E.P. and J.H.Th. Hengst, J. Colloid & Int. Sci., 29, 510 (1969)
64. Buchanan, A.S., and E. Heymann, Proc. Roy. Soc., (London) A195, 150 (1949)
65. Conwey, B.E., Electrochemical Data, Elsevier Pub., p217-221 (1952)
66. Overbeek, J.Th.G., Colloid Science, Ed. H. Kruyt, Vol. 1 Elsevier Pub. Co., Amsterdam, (1952)
67. Mukherjee, J.N., and J.K. Basu, Quart. J. Indian Chem. Soc., 3, 371 (1926)
68. Buzagh, A.V., and Z. Szaho, Kolloid Z., 83, 139 (1938)
69. Barr, J., and H.O. Dickinson, J. Photo. Sci., 9, 222 (1961)
70. Bovington, C.H., and A.L. Jones, Trans. Faraday Soc., 66, 764 (1970)

Substance	Growth Order	Supersaturation $(S-1)$	Ref.
AgCl	2	0.05-0.3	10
KCl	1	1.0×10^{-4} -1.5×10^{-3}	24
	1	1.0×10^{-2} -4.6×10^{-2}	25
NaCl	1		26
BaF_2	2.1	0.4-0.9	27
$BaSO_4$	3-1	20-62	28
	2	2-50	29
	2	0.5-10	30
	4-1	20-200	15
	3		31
	4		32
	2		33
	1		34
$PbSO_4$	2	0.05-0.4	35
Potash Alum	2		36
	1.6		37
$CuSO_4.5H_2O$	1.8	0.015-0.15	38
$MgC_2O_4.2H_2O$	2		40
$SrSO_4$	2	0.71-1.2	41
	3	7-24	16
$CaC_2O_4.H_2O$	2	0.86	42
	2	0.5-18	43
KH_2PO_4	2		44
$NH_4H_2PO_4$	2-1		44
$NaClO_3$	2-1		45,46
$Ba(NO_3)_2$	1.5-1	0.06-3.0	47
$K_2Cr_2O_7$	1.7	6.6×10^{-5} -6.6×10^{-4}	48
Ag_2CrO_4	2		49
	3		50
BaC_2O_4	2-1	0.01-0.5	46
$ZnC_2O_4.2H_2O$	2-1	0.1-23	51
$CaSO_4.2H_2O$	2	0.03-0.15	52
	1.5-2.1		53
	1.9		54
$CuSO_4$	1		55
	1		39
K_2SO_4	2	0.03-0.4	56
$BaCrO_4$	2		57
$CaCO_3$	2		58
	2	0.1-10	43
$PbCrO_4$	2	2×10^3 -2×10^4	59
$Ag_2C_2O_4$	2	0.4-3	43

Table 1. The dependence of growth rate on supersaturation for different substances.

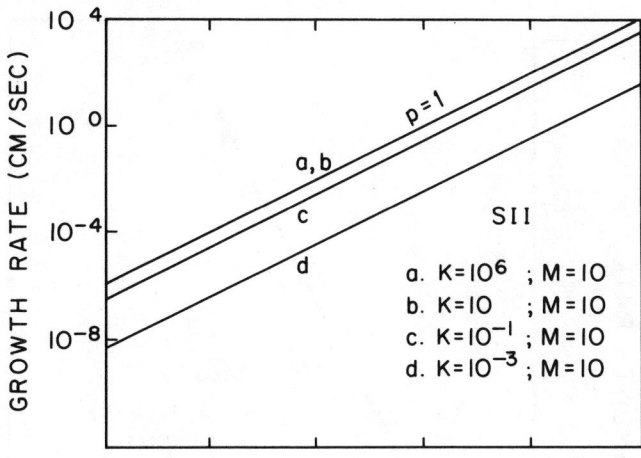

Figure 1a. The rate of growth versus supersaturation based on Mechanism II, sequential-ionic-integration, for a noncharged crystal, assuming a mass transfer coefficient of 10 cm/sec M^{-1} and varying the surface reaction constants from 10^3 to 10^{-3} cm/sec M^{-2}; (a) $K=10^3$, $M=10$, (b) $K=10$, $M=10$, (c) $K=1$, $M=10$, and (d) $K=10^{-3}$, $M=10$.

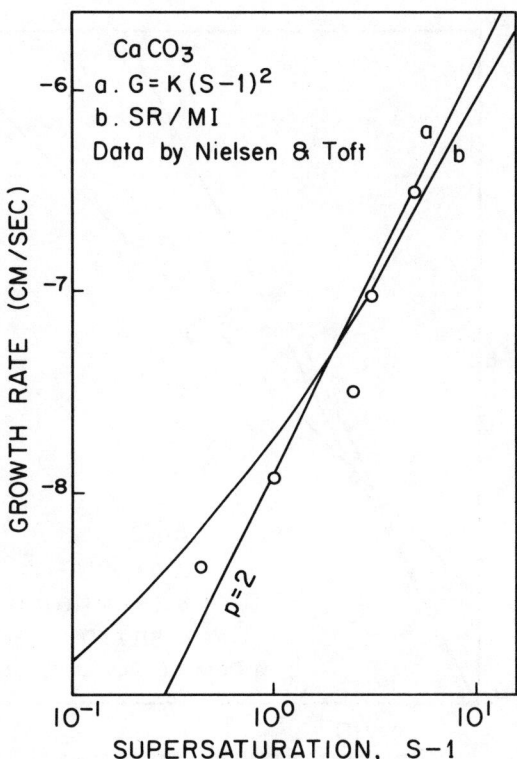

Figure 2. Dependence of the rate of growth upon supersaturation for calcium carbonate: a. obtained from second-order dependence, Equation 17; b. obtained from Equation 9, $G=K([A^+]_o^2-[A^+]_e^2)$; data taken by Nielsen and Toft (43).

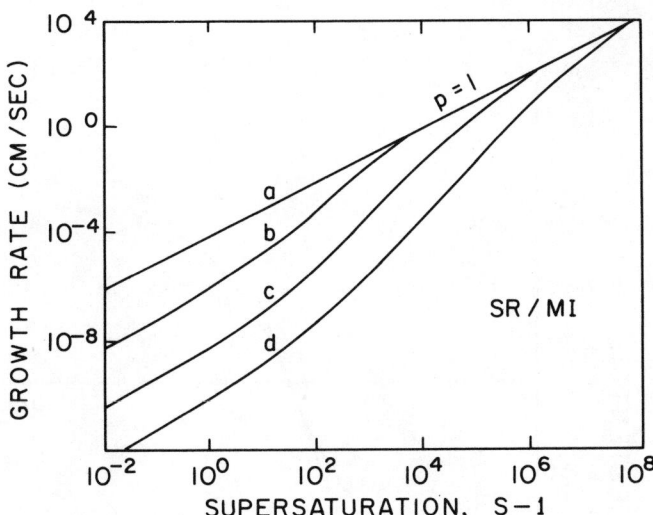

Figure 1b. The rate of growth versus supersaturation based on Mechanism I, surface-reaction/molecule-integration, for a noncharged crystal, assuming a mass transfer coefficient of 10 cm/sec M^{-1} and varying the surface reaction constants from 10^6 to 10^{-3} cm/sec M^{-2}; (a) $K=10^6$, $M=10$, (b) $K=10^3$, $M=10$, (c) $K=10$, $M=10$, and (d) $K=10^{-1}$, $M=10$.

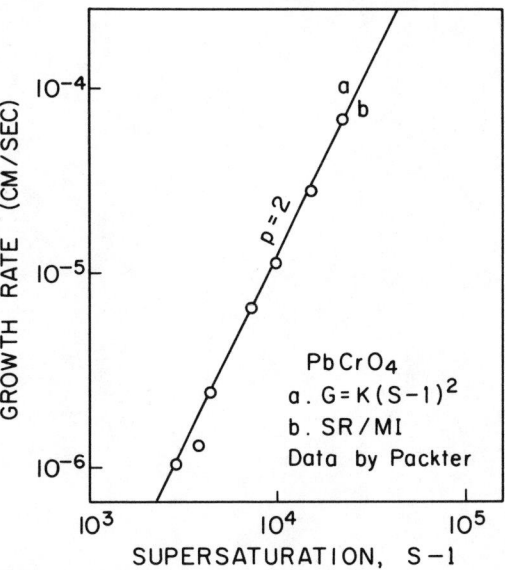

Figure 3. Dependence of the rate of growth upon supersaturation for lead chromate: a. obtained from second-order dependence, Equation 17; b. obtained from Equation 9, $G=K([A^+]_o^2-[A^+]_e^2)$; data taken by Packter (59).

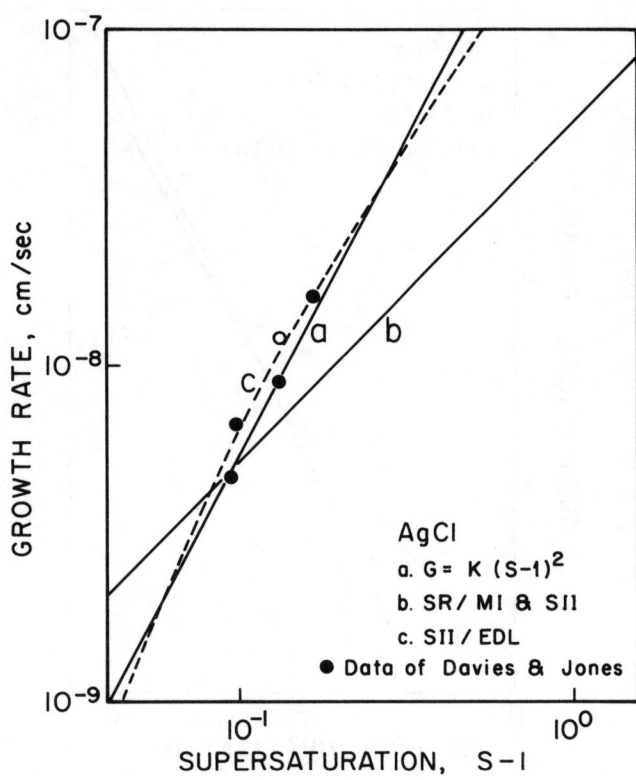

Figure 4. Dependence of the rate of growth upon supersaturation for silver chloride: a. obtained from a second-order dependence, $G=K([A^+]_o-[A^+]_e)^2$; b. obtained from $G=K([A^+]_o^2-[A^+]_e^2)$; c. obtained from the electrical double layer theory, Equation 16; data taken by Davies and Jones (10).

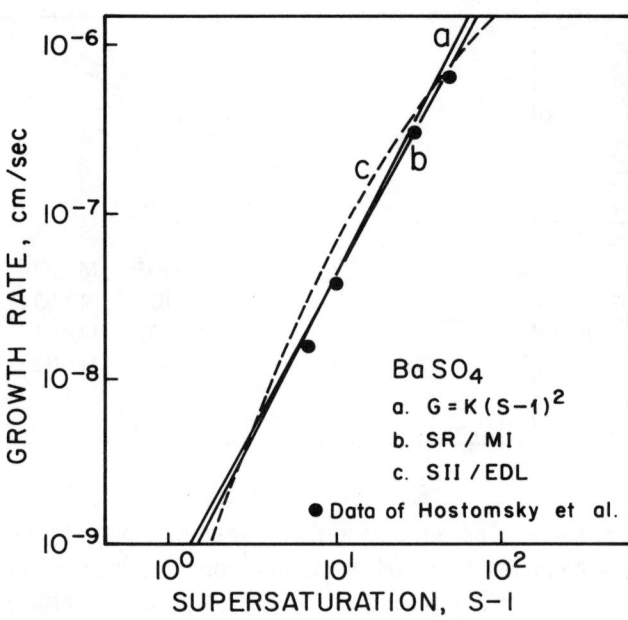

Figure 5. Dependence of the rate of growth on a wider supersaturation for barium sulphate: a. obtained from a second-order dependence, $G=K([A^+]_o-[A^+]_e)^2$; b. obtained from $G=K([A^+]_o^2-[A^+]_e^2)$; c. obtained from the electrical double layer theory, Equation 16; data taken by Hostomsky et al. (29).

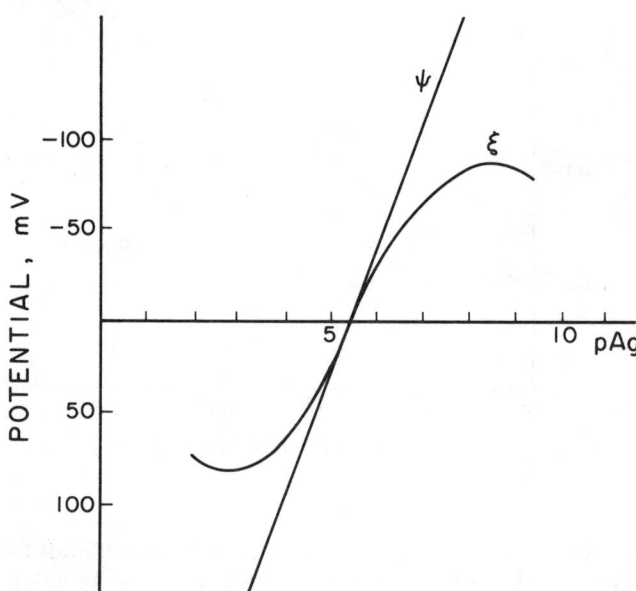

Figure 6. From Overbeek (66), surface potentials at a silver iodide/dilution interface as a function of the silver ion concentration expressed in pAg.

EFFECTS OF IMPURITIES ON THE PRODUCTION AND SURVIVAL STEPS OF CONTACT NUCLEATION

Gregory D. Botsaris and James O. Pagounes ■ Department of Chemical Engineering, Tufts University, Medford, MA 02155

Contact nucleation involves three steps: transfer of energy to a seed crystal, production of a number of potential nuclei, and survival of a fraction of them to become nuclei. This study was able to differentiate between the effect of impurities on the production and the survival steps. It was found that impurities affect the latter but not the former.

Secondary nucleation and in particular the so-called contact nucleation is the predominant type of nucleation in a large number of industrial crystallizations. Contact nucleation is the process by which a crystal seed breeds new nuclei by contact or collision with other solid surfaces. The understanding of the mechanism of secondary nucleation is essential for the development of a kinetic expression for the nucleation rate, which in turn is necessary for the design and scale-up of crystallizers.

The overall process of secondary nucleation is viewed as consisting of three stages, in series: (a) the formation of entities on the surface of the seed or adjacent to it, which can become potential nuclei, (b) the removal of such entities into the bulk solution by a collision mechanism, and (c) the survival of a certain fraction of these particles, that grow to macroscopic size.

In terms of this model, the nucleation rate N, can be expressed as the product of three functions;

$$\dot{N} = (\dot{E}_t)(F_1)(F_2) \qquad (1)$$

where, $\dot{E}_t$ is the rate of energy transfer to the crystals, F_1 is the number of particles produced per unit of collision energy, and F_2 is the fraction surviving to become nuclei.

The numerous investigations on the subject of secondary nucleation (see Reviews 1-4) are essentially attempting to clarify the above three functions, $\dot{E}_t$, F_1, and F_2, and study the effect of various variables on them. This investigation deals with one such variable, that is the impurities or additives invariably present in almost all crystallizations. Impurities have been shown to reduce the number of secondary nuclei produced (5-9). The question to be answered, however, is whether or not the impurities retard nucleation by affecting either F_1, or F_2, or both. In other words, do the impurities reduce the number of particles produced, by affecting the growth of the parent seed or do they reduce the number of particles surviving to become crystals? The approach and the experimental results reported in this paper are devoted towards clarifying this question.

APPROACH AND EXPERIMENTAL PROCEDURE

An experimental plan was developed, in which the growth of the seed crystal and the subsequent production of particles by contact (i.e. function F_1) could be isolated from the survival stage of the produced particles (i.e. function F_2).

The experimental apparatus consisted of a McCabe type contactor (see Figure 1), immersed in a nucleation vessel containing a supersaturated solution. In particular, a metal rod A was used to contact the crystal seed which was glued onto the platform C, using a minimum amount of EPOXY-5 resin and hardner. A solenoid-relay system provided on-off control of the contacting rod, i.e., support the rod up in on position, drop it when the relay is off. A spring was used to immediately raise the rod after contacting the crystal.

A tubular barrier B separated the nucleation vessel in two compartments, each containing a supersaturated solution with different impurity concentrations. The solution inside the barrier provided the environment for the growth of the seed crystal and the production of the potential nuclei through the contact of the seed with the rod. Immediately after the assigned number of contacts were completed the tubular barrier was removed, the two solutions were mixed by a stirrer and the resulting mixture provided the survival environment for the particles.

Sodium chlorate was used as the crystallizing species and sodium borate as the impurity. Four sets of experiments were performed at the experimental conditions shown in Table 1. In each set, environments (either production or survival) were created containing 5, 10 and 20 ppm of impurity.

The number of produced nuclei were counted two hours after contact, at which time a constant value had been reached.

RESULTS AND DISCUSSION

The data obtained from Set 1 are shown in Fig. 2. There, the average number of counted crystal ± the standard deviation is shown, as computed from four singular experiments performed for each point. The numbers indicate that the runs which had the same impurity concentration in the survival stage but different concentrations in the production stage produced numbers of nuclei which were not significantly different. In contrast, the numbers differed significantly in runs in which the impurity concentrations were the same in the production stage, but different in the survival stage. For instance, the numbers for the (5,5) and (10,10) experiments, 167 and 118, are about equal to those for (5,0) and (10,0) experiments 155 and 115. Therefore it can be concluded that the impurity did not affect the number of particles produced during contact. It affected, however, significantly the number of the particles which survived to become nuclei. In other words, the impurity affected function F_2 but not F_1.

This conclusion is supported also by the data in experimental sets II, III and IV (See Figures 3, 4, and 5). There are no significant changes in the vertical direction (i.e. increasing impurity concentration in the production stage). However, in the horizontal direction (i.e. increasing impurity concentration in the survival stage) the reductions of the nuclei number are considerable.

Experiments in Set II (Fig. 3) were performed to counter the possible argument that ten minutes was a too short time interval for the growth of the seed crystal and that the steady state growth configurations on the seed surface (which are expected to be different for different impurity concentrations in the solution) may have not been attained. This could be responsible for the invariability of the nuclei number in the vertical direction. The growth time increased to forty minutes. However, the trends remained the same.

The conclusion was also supported by the data in Set IV (See Fig. 5). There, not only the growth time but also the growth rate was increased by increasing the subcooling.

It should be noted, that while the introduction of the impurities in the production stage does not affect the number of particles produced by the contact, it does retard the growth rate of the seed crystal. This is shown in Fig. 6.

The experimental conditions of Set III differ from those of Set II only in the number of contacts, i.e. 2 instead of 5. An interesting point arises from a comparison of Figures 3 and 4, Fig. 4 indicates that for the (0,0) experiment the first two contacts created 767 nuclei while the five contacts created a total of 3215. This implies that later contacts were more productive. A possible explanation could be that after the first contact produces a certain number of particles, the subsequent contacts not only produce new nuclei, but also dislodge previously created and trapped nuclei.

CONCLUSION

In a secondary nucleation process impurities did not affect the number of particles produced during contact, but only the fraction of those particles which survived to become crystals.

LITERATURE CITED

1. Botsaris, G.D. in "Industrial Crystallization", Mullin, J.W. (ed) p. 3, Plenum Press, New York (1976).

2. Estrin, J., in "Preparation and Properties of Solid State Materials", Wilcox, W.R. (ed), 2, Marcel Dekker, New York (1976).

3. Garside, J. and Davey, R.J., Chem. Engng. Comm., 4, 393 (1980).

4. Larson, M.A., in "Industrial Crystallization 81", Jancic, S.J., de Jong, E.J. (eds) p. 55, North Holland, Amsterdam (1982).

5. Botsaris, G.D. and Sutwala, G., AIChE Symp. Ser., 72, 153 (1976).

6. Larson, M.A. and Khambaty S., Ind. Eng. Chem. Fundamentals, 17, 160 (1978).

7. Liu, Yih-An and Botsaris, G.D., AIChE J. 19, 510 (1973).

8. Larson, M.A. and Mullin, J.W., J. Crystal Growth, 20, 183 (1973).

TABLE 1
Experimental Conditions for the Contact Experiments

	Set I	Set II	Set III	Set IV
Growth period (minutes)	10	40	40	40
Number of contacts	5	5	2	2
Subcooling (°C)	1	1	1	2

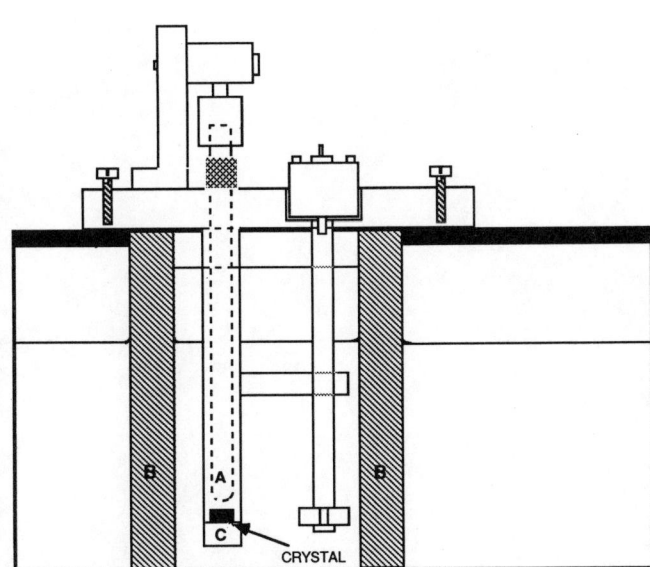

Figure 1. Experimental apparatus.

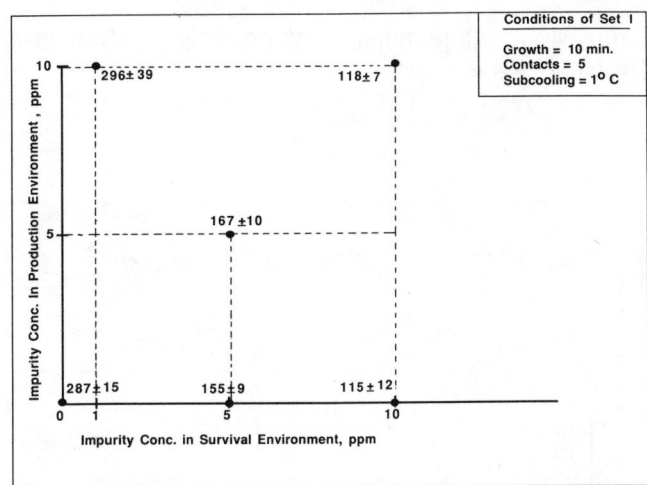

Figure 2. Average number of crystals (± Standard Deviation) Set I.

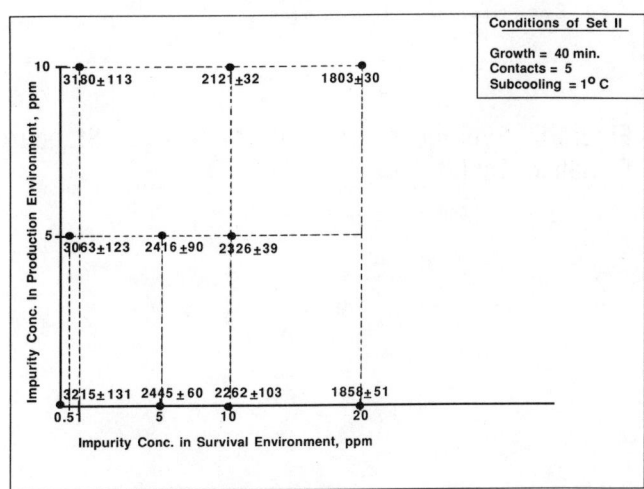

Figure 3. Average number of crystals (± Standard Deviation) Set II.

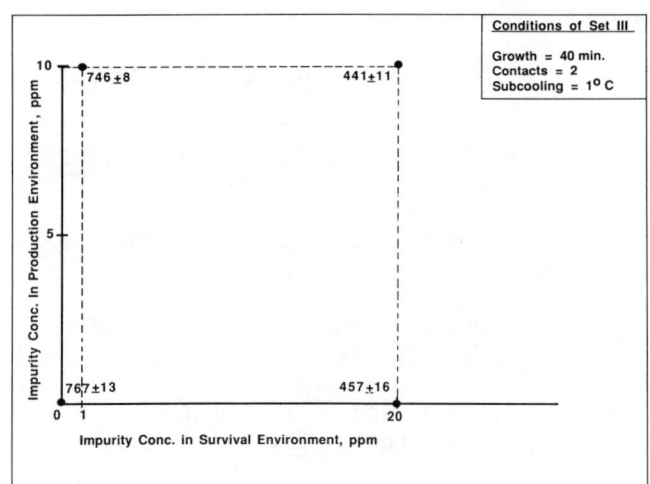

Figure 4. Average number of crystals (± Standard Deviation) Set III.

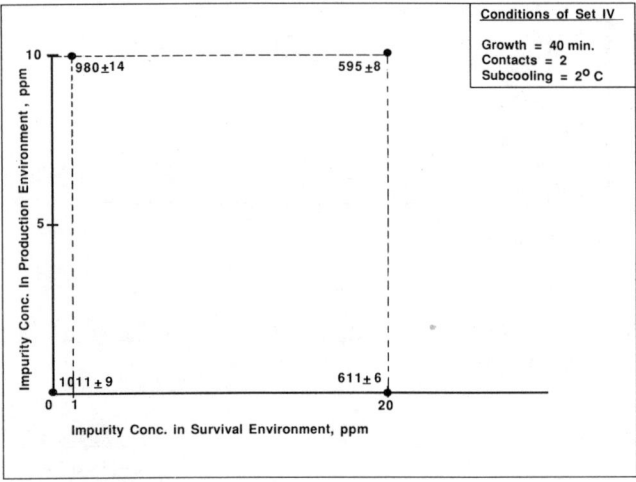

Figure 5. Average number of crystals (± Standard Deviation) Set IV.

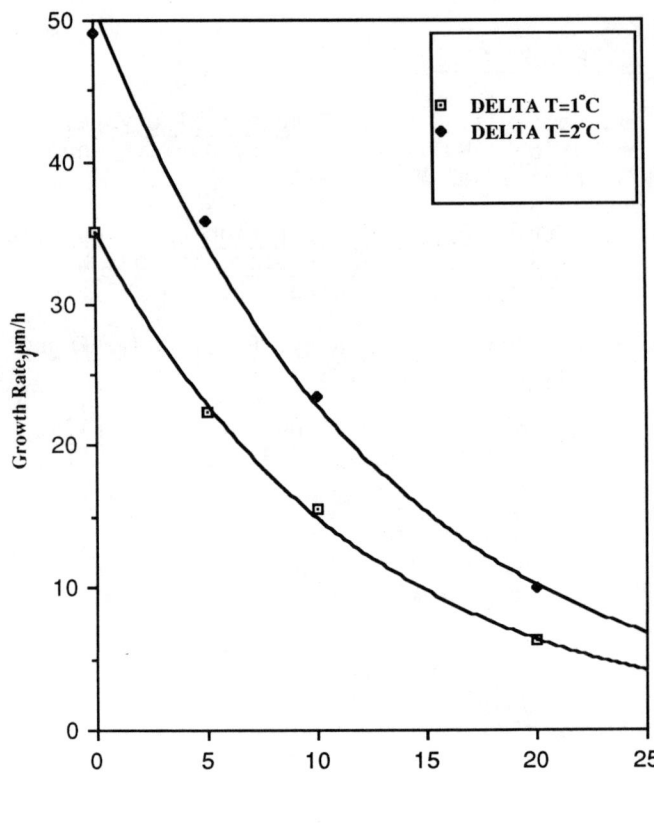

Figure 6. Growth rates of $NaClO_3$ crystals in the presence of Borax.

THE GROWTH OF GYPSUM

Walter F. Klima and George H. Nancollas ■ State University of New York at Buffalo, Department of Chemistry, Acheson Hall, Buffalo, NY 14214

The kinetics of crystal growth of calcium sulfate dihydrate has been studied microscopically for single crystals and by the constant composition method for well characterized seed suspensions. Good agreement was obtained between the linear growth rates calculated for each method and taking into account the rates of growth of individual faces. Experiments were made over a range of supersaturation and stirring dynamics and the results point to a spiral growth mechanism. Changes in growth rate are discussed in terms of decreasing spiral cooperation and bunching of steps on crystal surfaces. With decreasing crystal size, evidence is advanced for a change in mechanism from surface to diffusion control. Marked aggregation, crystal fracture and the development of secondary nuclei, monitored microscopically and by particle size distribution measurements, all influence the growth rates.

INTRODUCTION

The growth of calcium sulfate dihydrate, gypsum, is of considerable importance since it is a common component of scale which forms in a variety of industrial applications including desalination (1), oil production (2), and in cooling-tower technology (3). In phosphoric acid production, the precipitation of gypsum must be achieved with the minimum occluded produced acid. This is made difficult by a calcium sulfate hydrates solubility phase diagram markedly dependent upon temperature, ionic strength, and the presence of phosphate. Moreover, it has been shown that calcium sulfate dihydrate may nucleate and grow both on the surfaces of other minerals (4) and in the favorable temperature gradients at cooling-tower surfaces (5). In many of these processes, numerous small gypsum crystals of different shapes and sizes are formed with morphologies markedly dependent upon the presence of foreign ions in the solutions. Despite the importance of the gypsum system, the mechanism of the crystallization is still open to question although the rate of dissolution has been shown to be transport controlled (6,7).

Although numerous studies have been made of the rates of spontaneous precipitation of calcium sulfate dihydrate, the initial stages of the reaction are difficult to investigate owing to the problems associated with the detection of small nuclei. It is therefore difficult to design experiments to critically examine the mechanism of crystallization. In order to avoid many of the chance nucleation problems associated with these events, a highly reproducible crystal growth procedure was developed (8). The addition of well characterized seed crystals to stable supersaturated solutions of calcium sulfate enabled the rate of growth to be investigated by following concentration changes as a function of time. The rate of crystallization was shown to be proportional to the square of the supersaturation (Equation 1):-

$$dn/dt = ks \left\{ [(Ca^{2+})(SO_4^{2-})]^{1/2} - K_{so}^{1/2} \right\}^2 \quad (1)$$

In Equation 1, n is the number of moles of calcium sulfate precipitated in time t, k is the rate constant for crystallization, s is a function of the number of active growth sites, and K_{so} the gypsum solubility product. It was shown that the rate equation could be expressed either in terms of molar concentrations or activities of the lattice ions (8). Seeded crystallization experiments were made by Smith and Sweet (9) who showed that after an initial surge, representing approximately 30% of the reaction, the rate of crystallization could be expressed by an equation similar to (1) with s = $(m/m_0)^{2/3}$. Here, m_0 and m are the masses of crystals present initially and at time t, respectively.

An advantage of utilizing the seeded techniques is that crystal growth occurs on well defined surfaces of known area and morphology.

In the field, since precipitation invariably takes place on a surface already present, the seeded growth methods simulate field conditions much more closely than do spontaneous precipitation studies. The disadvantage of the conventional seeded growth experiments described above is the difficulty of determining the stoichiometry of the precipitates due to the appreciable reduction in concentrations of lattice ions during the reactions. Since the supersaturation decreases to zero as equilibrium is approached, analytical limitations preclude the precise determination of growth rates. In a number of studies, relatively high supersaturations were used in an attempt to improve the statistical reliability of the results. However, this can result in a change in the mechanism of the precipitation processes.

These problems were overcome in the constant composition method, first used for the study of the crystallization of calcium phosphates (10). It was later extended to calcium sulfate dihydrate using a calcium specific ion electrode (11) and a conductometric technique (12) for controlling the supersaturation. The method is particularly useful for investigating the mechanism of crystal growth, since the extent of the reaction at any chosen level of supersaturation can be varied widely. In the present work, the constant composition method has been used to study the kinetics of crystallization of gypsum in supersaturated solutions over a range of concentrations and fluid dynamics. In a parallel study, optical microscopy has been used to measure the linear growth rate of well formed seed crystals under similar concentration conditions.

EXPERIMENTAL

Triply distilled, deionized, carbon dioxide-free water was used throughout and the constant composition (CC) crystallization experiments were conducted in a nitrogen atmosphere. Reagent grade chemicals were used and calcium chloride and sodium sulfate were recrystallized twice from aqueous solution. Stock solutions were filtered through Millipore filters (0.22 μm) which were pre-washed with water. Calcium specific ion electrodes (Radiometer) were standardized in calcium chloride solutions of known ionic strength. For the control of pCa, a Metrohm potentiometer was used in combination with a Metrohm Impulsomat and recording Dosimat. In order to maintain the composition of the supersaturated solutions at known values during the CC crystal growth reactions, the Dosimat was modified to deliver two solutions simultaneously from coupled burets. Calcium analyses were made by EDTA titration using murexide as indicator, by ion exchange, or by atomic absorption spectrophotometry (Perkin Elmer Model 503). Solid phases were studied by scanning electron microscopy (ISI Model II) and confirmed as gypsum by x-ray powder diffraction (Phillips XRG 300 diffractometer, CuKα radiation, Ni filter) infrared spectroscopy (Perkin Elmer Model 521). The particle size distribution during the crystallization reactions was monitored with an Electrozone/Celloscope particle counter (Particle Data Inc.) by suspending samples in saturated solutions at the same ionic strength to provide a suitable conducting medium. Specific surface areas were measured by a single point BET method using 30/70 nitrogen/helium gas mixture (Quantasorb, Quantachrome, Inc.).

Large crystals of good morphology were grown in supersaturated solutions (50 mL) over which heated water-saturated nitrogen was passed. The needle shaped crystals nucleated at the hot solution/gas interface and grew relatively free of macroscopic steps and indentations to sizes ranging from 100 to 1,000 μm. These crystals were used in the single crystal optical microscopy growth experiments. Seed suspensions for CC crystallization were prepared by the slow mixing of calcium chloride and sodium sulfate with the appropriate background electrolyte (normally 0.2 mol L^{-1} sodium chloride) at 25 °C. After one hour, crystals of 40-50 μm were separated by sedimentation and incubated for two hours in 0.035 mol L^{-1} calcium chloride solution in order to improve their surface structure. They were aged and stored at 25 °C in saturated calcium sulfate solution, 0.2 mol L^{-1} in sodium chloride.

Crystal Growth Experiments:

The morphology of the growing crystals was studied in situ by optical microscopy (Unitron microscope) and by scanning electron microscopy of the filtered solids. Single crystal growth studies were made at 25$\pm$0.01 °C using a thermostatted cell (150 mL) adapted for use on a microscope stage. Linear growth rates were calculated from the results of at least 50 crystal measurements. The supersaturated solutions (0.035-0.055 mol L^{-1} calcium sulfate/0.2 mol L^{-1} sodium chloride) were stirred with an overhead magnetic bar spun at a known speed using the magnetic field of another magnet attached to the shaft of an adjacent variable speed motor. The cell was covered with a Plexiglass lid and flushed with water-saturated nitrogen during experiments. 1 to 10 crystals were added to the solution and changes in length, width and thickness dimensions of the crystals were monitored. The supersaturation was effectively constant during growth since the amount of precipitation was a negligible fraction of the available calcium sulfate present. In the single crystal studies, the optical microscope was used to monitor the morphology and size of the fragments and the strength of the crystals was determined by positioning them lengthwise over a small slit on a stationary platform and applying weights to a wire (1 mm diameter) looped around the width of the crystals.

Following the addition of seed crystals in the CC experiments the supersaturation was maintained constant by the potentiostatically controlled addition of calcium chloride and sodium sulfate using the specific ion calcium electrode. The two titrants normally consisted of 0.100 mol L^{-1} calcium chloride and 0.100 mol L^{-1} sodium sulfate, each 0.2 mol L^{-1} in sodium chloride in order to avoid changes in ionic strength. The rate of growth was calculated from the rate of titrant addition after corrections for dilution. In order to verify constant solution concentration during the experiments, aliquots were filtered through 0.22 μm Millipore filters and analyzed for calcium ion. Particle size distribution measurements were also made during the crystallization experiments.

Secondary nucleation was investigated in the thermostatted cell both under constant composition and optical monitoring conditions. In the former case, the crystals were either suspended freely in solution or else mounted with epoxy onto the end of the glass rod. A vibrational shock could be transmitted to the mounted crystals by tapping. The overhead mixing impeller was positioned high in the reaction cell in order to avoid impeller-crystal collisions.

RESULTS & DISCUSSIONS

Crystal Growth:

A typical plot of average linear growth rates, $\dot{l}_{avg}$, measured over the lengths of crystals having initial lengths, $l_o \geq 100 \mu$m, as a function of calcium sulfate concentration is shown in Fig.1. Similar plots were obtained for the rates of change of average crystal width, $\dot{w}_{avg}$. The linear crystal thickness growth rates $\dot{w}'$, were assumed to be constant since they were some thirty times less than $\dot{l}_{avg}$. For crystals smaller than 100 μm, $\dot{l}$ was dependent upon the initial crystal length l_o as shown in Fig.2. Above 100 μm, however, $\dot{l}$ was constant ($\pm$10%). At high supersaturation (0.052 mol L^{-1} calcium sulfate/0.2 mol L^{-1} sodium chloride) values of $\dot{l}$ were less than $\dot{l}_{avg}$ at stirring speeds below 30 rpm. At higher speeds (30-200 rpm), however, the $\dot{l}$ values were within 10% of $\dot{l}_{avg}$. Measurements reported here were made at stirring speeds 40-50 rpm, a range in which no effect was observed on the linear rate of growth. Higher stirring rates tended to reorientate the crystals in the microscopic cell and suspend them in the solution (70-100 rpm). Measured linear growth rates in Fig.1 were reproducible to within $\pm$ 15%.

Linear growth rates were also measured for smaller crystals ($\leq$ 20 μm) which were aggregated either to the edges or faces of crystals larger than 100 μm in length. The growth rates in 0.049 mol L^{-1} calcium sulfate/0.2 mol L^{-1} sodium chloride solution were higher (0.07 mmh^{-1}) for crystals aggregated to the edges of larger crystals than when they were free in solution (0.01 mmh^{-1}). However, when they were aggregated to the faces of the larger crystals, the effect was not as great (0.02-0.03 mmh^{-1}). The average growth rates, $\dot{l}_{avg}$, for the small crystals which were part of clusters of similarly sized particles (0.08 mmh^{-1}) were also higher than for the freely suspended crystals.

The results of typical constant composition experiments are shown in Fig.3 from which it can be seen that the growth rate was constant at each supersaturation. Changes in total crystal volume as a function of time, $v(t)$, were calculated from the rates of titrant addition in Fig.3 and expressed in terms of the increase in volume for the known number of crystals. Using the results of the optical microscopy experiments for $\dot{w}$ ($\approx$ 0.1 l mm h^{-1}) and $\dot{w}'$ ($\sim$0.03 l mm h^{-1}) the calculated values of $\dot{l}_{avg}$ were within 25% of those determined from direct microscopic observations.

Linear growth rates obtained from single crystal growth experiments showed relatively constant values for the ratio of the rates of increase in crystal thickness to that of width ($\dot{w}'/\dot{w}$ = 0.33 $\pm$ 0.1) and for the ratio of the growth rates of crystal length to crystal width ($\dot{l}/\dot{w}$ = 10). The initial dimensions of length/width, l_o/w_o varied between 1 and 20 indicating that the rate of growth changed during the development of the seed crystals. As the crystals grew larger, they tended to fracture along the length axis which would explain why the crystals grew at about the same rates despite their varied shapes. Since the ratio of initial thickness/width (w'_o/w_o) remained relatively constant, the volume of the crystals could be estimated from the measured values of l and w by accounting for the geometric shape of the crystals. Resulting plots of ln $v(t)/v(o)$ against ln σ are shown in Fig.4. The calculation of the normalized volume change assumed the same initial crystal volume, $v(o)$ in each of the supersaturated solutions. In Fig.4, the relative supersaturation, σ, is given by

Equation 2:-

$$\sigma = \left\{ [(Ca^{2+}(SO_4^{2-})]^{1/2} - K_{SO}^{1/2} \right\} / K_{SO}^{1/2} \qquad (2)$$

The slope of the straight line, 2.0 ±0.07, confirms the parabolic rate law (Equation 1) for both the microscopic and CC experimental results. In Fig.4, plots are also included for growth results predicted for initial volumes increasing by factors of 1.3, 1.6, 1.9, and 2.2 . It can be seen that the corresponding plots form a series of approximately parallel straight lines of similar slope but with separation between successive lines (d in Fig.4) decreasing as the size of the crystals increased. Since the linear growth rates are relatively constant (±10%) for crystals larger than about 100 μm at each supersaturation, changes in crystal volume will be proportionately less as given by the ratio $\dot{v}(t)/v(o)$, bringing the successive lines closer together. The results in Fig.4 are in good agreement with those of van Rosmalen et.al.(12) who studied the crystallization of gypsum using a conductometric monitoring technique. Plots of ln $\dot{v}(t)/v(o)$, also showed a series of parallel straight lines with separation distance between successive lines decreasing as the number of crystals was increased (12). However, in the latter study, the number and size of the crystals were less defined than in the present work in which the reference line (0-0-0) in Fig.4 explicitly refers to a crystal of defined dimensions and volume. Calculations involving the results of seeded growth experiments require the dissipation of mass uniformly over each crystal and subsequently each face or through a geometric factor to fit the results. In either case this would introduce serious errors since the linear growth rates have been shown, in the present work, to differ along the various crystal dimensions (l ≠ w ≠ w'). Qualitatively, however, the position of van Rosmalen's data in Fig.4 corresponds to a lower ordinate intercept reflecting the use of larger sized crystals.

The observed independence of linear growth rates upon crystal size (≥ 100 μm) may result either from step bunching or cooperation between spiral centers. If the step velocities are not uniform, faster growing steps may bunch up behind slower moving ones, decreasing the growth for fast moving steps and leading to the observed reduction in the overall growth rate. In general, macroscopic step bunching would be expected to be more common on crystals of larger size since the probability is greater for steps to interact. This was observed during the early stage of the crystallization and the steps were observed to move across the crystal faces with faster moving steps catching up with slower ones. In the later stages of the crystallization, the smooth (010) faces were relatively free of macroscopic steps (500X). However, after the addition of sodium hexametaphosphate, many steps appeared on the smooth crystal surfaces suggesting that the inhibitor molecules decreased the growth rates causing the less inhibited steps to bunch together.

Spiral dislocations on a crystal surface may interact if their centers are within a certain distance. If the centers are of the same sign, the rate of growth may increase. However, if they move apart as the crystal grows, the extent of cooperation would be decreased. The kinetic results of this work fit a parabolic rate equation which may be the result of a spiral growth mechanism. Tetragonal spirals were observed on the (010) faces of large crystals during dissolution in stagnant solution. This phenomenon was also observed by van Rosmalen et.al. (13) during gypsum dissolution in a gel. In the dissolution experiments, some of the etch pits were visible for long times (primary) while others were short lived (secondary). Primary etch pits may have resulted from dissolution along the dislocation lines of spiral centers which had developed during growth. It is interesting to note that some of the etch pits overlapped in a manner suggesting cooperation between spirals.

Aggregation:

It can be seen in Fig.5, A, B and C, that in the early stages of the CC experiments, the rate of growth decreased as the number of crystals and specific surface area decreased. These results suggest that crystal aggregation resulted in a decrease in the rate of crystallization. Linear growth rate experiments showed that the fractured seed crystals grew rapidly at rates greater than l_{avg} for short periods of time (~10 min) followed by normal growth (~l_{avg}). This process of morphology perfection may therefore also have been partly responsible for the reduction in growth rate.

The addition of insoluble metaphosphate (IMP) in lithium chloride solution to saturated calcium sulfate solution markedly increased the rate of gypsum crystal aggregation as measured by particle size distribution. The observed aggregation rate closely followed the Smoluchowski equation (14,15), as shown in Fig.6. Increasing either the size or concentration of seed crystals resulted in higher rates of aggregation, with coagulation times varying inversely with the initial number of seed crystals. Aggregate size also increased at higher concentrations of IMP but at the higher stirring rates, only crystals joined through flat (010) faces remained together. These large flat surfaces provided the maximum area of contact between crystals with the resulting strongest binding interaction. In experiments involving a wide range of crystal sizes, small crystals tended to cluster close to the center of the aggregates with the larger crystals extending into the solution. Very small crystals (~0.1 μm) formed spherical and densely packed aggregates.

In the presence of IMP, aggregation increased greatly with stirrer speed probably resulting from the bridging of adsorbed polymer IMP molecules (16). Although the rate of gypsum aggregation followed the Smoluchowski equation, the time of aggregation was much less than that predicted for spherical particles. This may be attributed to the orthokinetic effect caused by the combination of large crystal size and the stirring of the solution. Needle-shaped gypsum crystals have a much larger sphere of attraction as compared with spherical particles of equal mass (17). Small-large crystal collisions will be more frequent than those involving only small particles. This may explain the clustering of small nuclei on the surfaces of parent crystals during secondary nucleation and the agglomeration of small crystals to large crystals during crystallization. It is interesting that agglomeration tended to occur on the flat (010) faces with the largest available contact area.

Crystal Fracture:

One of the most difficult problems in attempting to understand the mechanism of crystallization of gypsum, is the tendency of the crystals to fracture during collisions between crystals, with stirrers or with walls of the reaction vessels. Small flaws in the solid structure develop under local stress and become visible as cracks. Transmission electron microscopic studies of Hockey and Lawn indicated that the atoms lining the sides of a crack can realign and bind together (18). It was suggested that the crack behaves as an atomically sharp slit which may close and heal as easily as it may open and grow (the zipper effect). Thus when a bond at a crack tip is stretched beyond a certain distance, it breaks, displacing the cracked tip. This is opposed by the retarding force due to the constraints of the linear elastic lattice surrounding the cracked tip and the restoring force of the stretched bond. In terms of this mechanism, the degree of gypsum fragmentation depends upon the amount of energy transmitted to the crystals. This is influenced primarily by the characteristics of the impeller (size, shape, rotation speed, and composition) and the intrinsic strength of the gypsum crystal structure which may vary with size, shape, method of seed preparation, and the presence of impurities.

In this study, fragmentation was observed to occur along the planes of lattice ions which separate double layers of water molecules in the parent crystals. In contrast to gentle nitrogen bubbling and the use of soft polycarbonate stirrer materials, stainless steel impellers greatly increased the degree of fragmentation as did stirrers which were flat as opposed to being round in shape. Magnetic stirring produced a large number of fragments of wide size range and the seed crystals were greatly distorted in shape. Higher stirring rates increased the rate of fragmentation and decreased both the size and size range of the fragments. Seed crystals removed from suspension and air-dried were appreciably weaker than crystals which were kept in solution. However, aged and rapidly grown crystals cleaved under considerably less weight than those which were pre-grown by constant composition. Dendrite tips (outer 0.1 - 1 μm) were especially susceptible to fracture. However, the remainder of the dendrite was not easily broken even when the crystal was agitated against the cell walls.

In the fracturing process, the applied stress may rupture bonds along a plane forming either a crack or a shift in lattice ion layers in which new bonds are formed to neighboring atoms. Gypsum is very susceptible to cleavage along double layers of water molecules which separate adjacent lattice ion planes. During fracture, many parallel cracks first form along these planes with the largest probably originating from the point of impact and decreasing in size as the energy dissipates along the length of the crystal. Formation of abundant cracks on large crystals suggests that single collisions are often of insufficient energy to cause cleavage. However, the formation of cracks is important since the energies from the relatively weak collisions can accumulate, resulting in crystal cleavage. Not only may fracture of crystals increase the rate of crystallization due to the production of additional particles, but the process leaves rough edges at which accelerated growth may take place. It is therefore important during the kinetics studies to keep fragmentation to a minimum.

Secondary Nucleation:

Constant composition experiments were made using crystals which were either pre-grown (by CC) or aged in saturated solution for one month. Their stability was verified by exposure to water for about 1 sec; well formed crystals dissolved at edges whereas aggregated crystals disintegrated. To eliminate small particles which could act as potential nuclei, the solutions were filtered 3 to 5 times through .22 μm Millipore filters until only background noise was recorded on the particle data instrument.

Single crystal experiments were made either suspended in solution or else mounted with epoxy cement onto the end of a glass rod through which vibrational shocks could be transmitted. In a typical CC experiment at higher supersaturation, the rate of growth markedly increased as the reaction progressed. During this period, many of the crystal tips became broken producing secondary nuclei in numbers increasing with supersaturation. Increasing the rate of stirring also induced more crystal-wall collisions forming a greater number of small fragments. Typical size distribution curves are shown in Fig.7. In some experiments at lower supersaturation, the number of small fragments decreased with time due to the formation of clusters 8-10 μm in size. In the early stages of crystallization, many nuclei became preferentially attached to the faces of the parent crystals but were relatively mobile on the surface and were directed by the flow of solution to cluster at the downstream edge of the larger crystals. When the parent crystals were either tapped or agitated against the cell walls, the clusters dislodged from the surfaces and became new secondary nuclei. The process was quantified by recording the number of new crystals < 2 μm in size. Linear growth rates of the crystals decreased with decreasing size. Although many of the produced nuclei should therefore have shown negligible growth rates, the rapid clustering together produced larger bodied aggregates which grew at appreciable rates.

ACKNOWLEDGMENTS

We thank the National Science Foundation for a grant (# CPE 831338301) in support of this research.

LITERATURE CITED

1. Stumm, S. and Morgan, J.J., Aquatic Chemistry, Wiley-Interscience, New York, 1970.

2. Vetter, O.J., J. Pet. Tech., 1299 (1970).

3. Cowan, J.C. and Weintritt, D.J., Water-formed Scale Deposits, Gulf Publ. Co., Houston, 1976.

4. Gill, J.S. and Nancollas, G.H., Desalination, 29, 247 (1979).

5. Nancollas, G.H. and Klima, W.F., Mat. Performance, 21, 9 (1982).

6. Liu, S.T. and Nancollas, G.H., J. Inorg. Nucl. Chem., 33, 2311 (1971).

7. Christoffersen, J. and Christoffersen, M.R., J. Crystal Growth, 35, 79 (1976).

8. Liu, S.T. and Nancollas, G.H., J. Crystal Growth, 6, 281 (1970).

9. Smith, B.R. and Sweet, F., J. Colloid Interface Sci., 37, 612 (1971).

10. Tomson, M. and Nancollas, G.H., Science, 200, 1059 (1978).

11. Kazmierczak, T. and Schuttringer, E, Tomazic, B. and Nancollas, G.H., Croat. Chem. Acta., 54, 277 (1981).

12. van Rosmalen, G.M, Daudey, P.J. and Marchee, W.G.J., J. Crystal Growth, 52, 801 (1981).

13. van Rosmalen, G.M, Marchee, W.G.J. and Bennema, P., J. Crystal Growth, 35, 169 (1976).

14. von Smoluchowski, M., Z. Physik, Chem., 92, 129 (1917).

15. Kruyt, H.R., "Colloid Science", Elsevier, New York, 1952.

16. Lyklema, J., Proc. 5th Int. Cong. Surface Activity, B, 134 (1968).

17. Stober, W., Reinhalt Luft., 30, 277 (1970).

18. Hockey, B.J. and Lawn, B.R., Mat. Sci., 10, 1275 (1975).

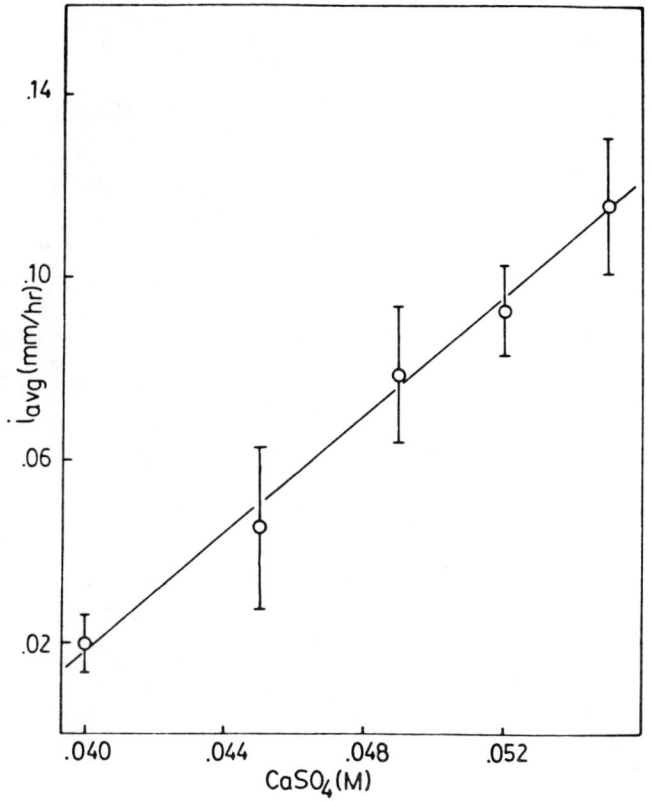

Figure 1. Average linear growth rate plotted as a function of supersaturation expressed as molar calcium sulfate (M).

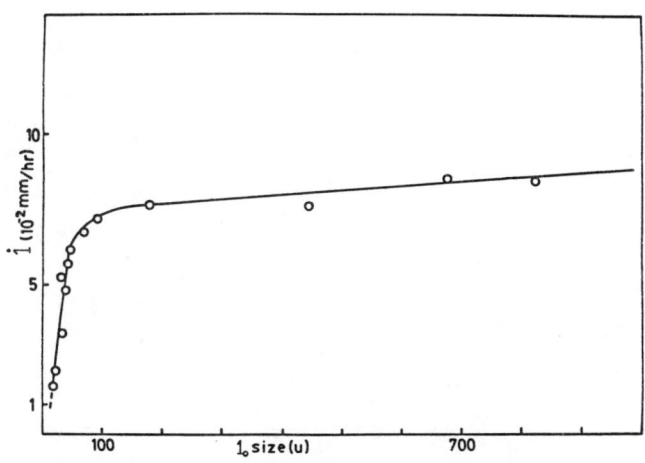

Figure 2. Linear growth rate 1 as a function of initial crystal size l_0.

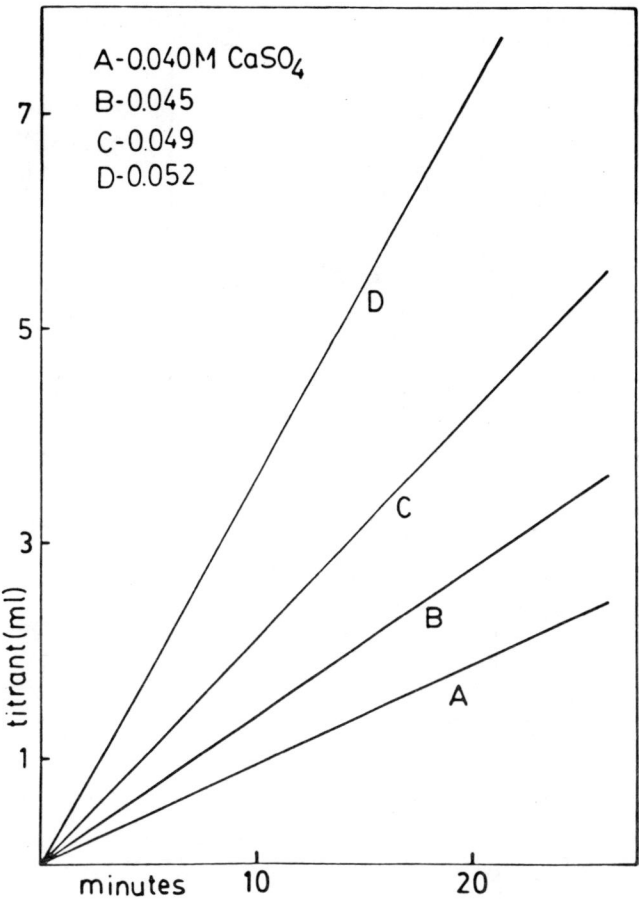

Figure 3. Plots of titrant addition as a function of time in the CC growth of gypsum.

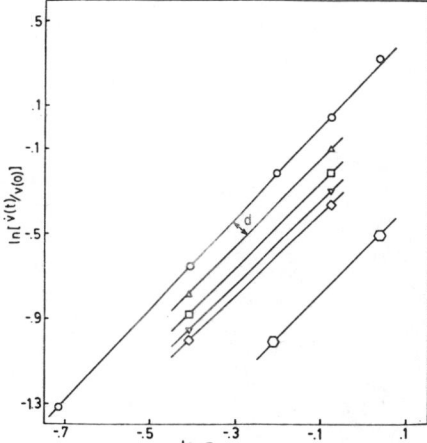

Figure 4. Plots of $\ln \dot{v}(t)/v(o)$ against $\ln \sigma$. Growth of crystal of initial volume, $v(o) = 10^{-4}$ mm^3, ○. Calculated rate data for crystals of volumes increasing by factors of 1.3, (△); 1.6 (□); 1.9 (▽) and 2.2 (◇) as compared with $v(o)$. Data from ref. 12, ⟡.

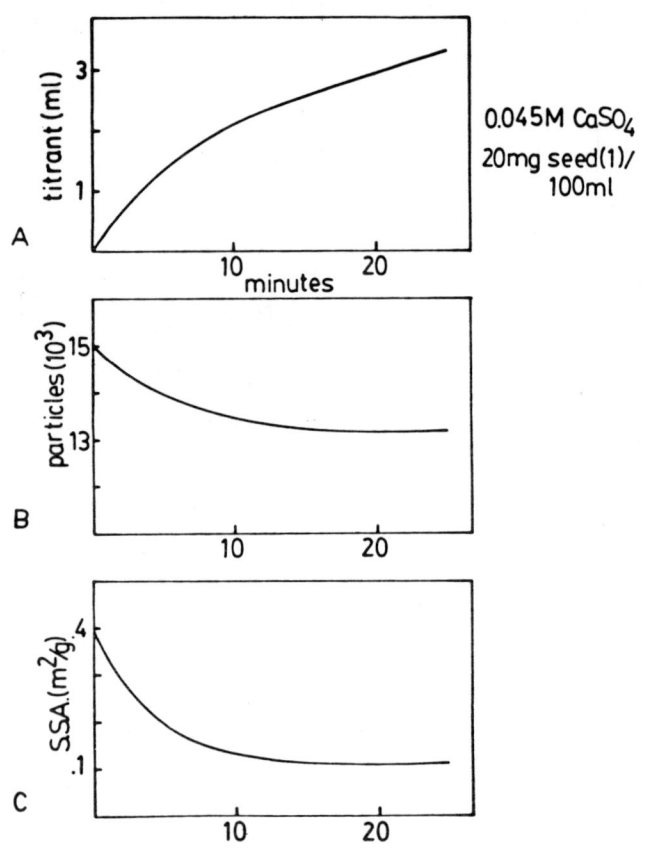

Figure 5. Aggregation of gypsum crystals during CC growth as reflected by particle sizes and specific surface areas.

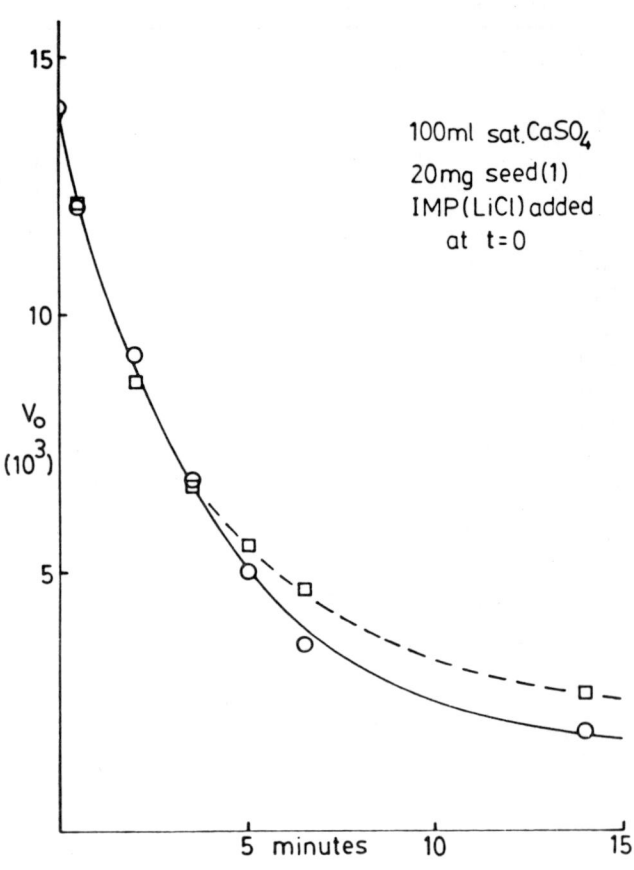

Figure 6. Total number of particles as a function of time during aggregation, ○ • □, as predicted by the Smoluchowski equation.

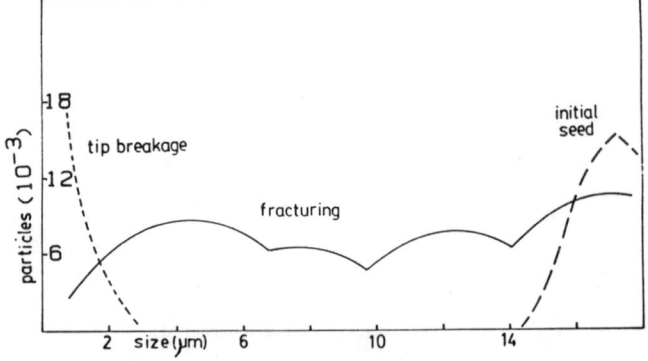

Figure 7. Size distributions of gypsum crystallites at various levels of fragmentation.

WAX HABIT IN CHILLED DIESEL FUEL, A GROWTH MODEL

E. Halter, J. Kanel, and D. Schruben ■ Department of Chemical Engineering, The University of Akron
Akron, OH 44325

The influence of composition and cooling rate on the size of paraffin (n-alkane) wax solids in some chilled synthetic and petroleum products has been well noted, but actual growth rate of n-alkane crystals has never been measured. Experimental measurements of growth rates resulted in the detection of growth rate dispersion in this system.

INTRODUCTION

Paraffin waxes appear in nature a variety of ways, for example in the waxy coatings on fruits, or in petroleum. It is not surprising, then, that they have been extensively studied (Turner, 1971). The terms paraffin and normal alkanes sometimes are applied interchangeably, however, paraffin waxes usually connote a material with a variety of molecular weight n-alkanes, and perhaps impurities or other aliphatic isomers while the term normal alkanes is reserved for fairly pure forms of normal alkanes. It is in the later sense that both terms are usually considered in studies of petroleum fuels, shale oils and to a lesser extent, coal liquids or other synthetic fuels (Branthaver et al., 1983). Thus, the terms are equivalent herein.

Precipitated paraffin waxes in many petroleum products can cause drastic change in their quality. This point of concern has been the springboard for many inquiries (McMillan et al., 1981) (Steere et al., 1981) into wax precipitate crystal size as well as shape. The influence of cooling rate on wax crystal size has been noted particularly well. However, the actual crystal growth rate has never been measured, until this study. Such scientific endeavors with respect to waxes are appropriate, since waxes will continue to be an important aspect of evolving fuel stocks such as heavy oils and shale oils.

RESULTS AND DISCUSSION

Components such as isoparaffins can be influenced in the paraffin waxes that form in mid-distillate fuels at low temperatures and cause problems such as filter plugging. The dominant influence on fuel cold flow, though has been recognized to be the normal alkanes. These studies (McMillan et al., 1981) (Steere et al., 1981) have recognized the importance of crystal size and shape to field performance. They agree with work of perhaps a more physical chemistry flavor (Feldman, 1983)(Holder et al., 1965) that is interpreted to conclude as follows:

1. Size and shape can be influenced by composition as for example when additives diminish what would otherwise be the natural crystal size distribution in a fuel (McMillan et al., 1981). Another example is when presence of C_{20} n-alkane in a dilute

---

D. Schruben is currently at Texas A&I University, Kingsville, Texas 78363

C_{24} n-alkane solution in solvent causes smaller crystals to form (Holder et al., 1965)(Holder, 1966).

2. Ultimate crystal size (if not shape) is influenced by cooling rate (McMillan et al., 1981)(Steere et al., 1981).

It is unfortunate that size dependence on cooling rate has not been better quantified. Totally unquantified has been the conditions an n-alkane crystal goes through to reach a particular size, that is to say the growth habit.

The C_{24} n-alkane species was selected for study and it is worthwhile to see why this was done. Most of the background material with photographs of n-alkane crystals pertinent to this study comes from the fuel literature. One naturally would be interested then in what waxes are most important in, say, a diesel fuel. Perhaps those are the waxes that first start to solidify as a diesel fuel is chilled. Which n-alkanes solidify and precipitate depends on which ones were initially present in the fuel. Wax distributions in diesel fuel typically peak in the C_{15} to C_{20} range. However, a scan of current work (e.g. McMillan et al., 1981) shows that perhaps commonly the C_{24} n-alkane is the first to precipitate in recordable amounts (or at least representative of them). Hence, it was chosen for this study.

What has been accomplished here is quantitative measures of crystal growth over time for the C_{24} n-alkane crystals that form as a dilute heptane solution of that species is cooled. (The experimental details are discussed in the next section.) The result in Fig. 1 shows relative crystal size at ten second intervals from an arbitrary zero time well after growth has started. It should be emphasized that Fig. 1 is a quantitative statement about n-alkane crystal growth rates, as distinct from qualitative statements about crystal size dependence on cooling rate made in earlier studies.

The first conclusion drawn from Fig. 1 is that growth rate is constant for a given crystal. The linear regressions used to fit Fig. 1 data had correlation coefficients (r^2) of 0.98 or better. This is significant because size dependent growth (Abegg et al., 1968) was one of the most reasonable and studied assumptions in the relatively larger inorganic crystal dynamics literature (compared to n-alkane crystal dynamics literature).

Another major finding is that the slopes, that is to say, the growth rates are different (distributed). This could be established by considering a confidence interval on correlation slopes or by confidence intervals on observations, but the conclusion that the slopes are different appears obvious from Fig. 1, without sketching in some kind of a confidence interval, especially given the high r^2 values. Perhaps more severe is to inquire about a confidence interval for the slopes, at which we find the two highest slopes of the six shown lie outside of 95% confidence interval based on all six slopes. If the highest slopes are conjectured to be from a different population, then the two groups could be formed and a comparison of means made. Here again, the finding is that growth rates are different.

All the data in Fig. 1 are believed to have been taken under identical conditions. As noted in the experimental section, this includes careful control of cooling rate and concentration. Subtle possibilities of a secondary nature remain, such as growing crystals or nucleation in the vicinity of a slow growing crystal causing a local depression in supersaturation. To the best of our current observations and calculations, secondary effects are not active. Photograph sequences start in the early stages of crystallization where crystal growth is towards a large crystal-free volume. In cases where crystals are in close proximity, separations are on the order of millimeters. Liquid diffusion coefficients are too small to cause substantial mass transport on the 10 second time scales involved here. Since all the data in Fig. 1 were taken under identical conditions, what must be concluded, at least tentatively, is that these organic crystals (C_{24} n-alkanes) display growth rate dispersion. This is significant, because this may have been only first discovered for the more studied inorganic class of systems as recently as 1981 (Bergland, 1981). It seems remarkable that this observation should not have been reported in the interval from well before Louis Pasteur's time to so recently, and so the conjecture that it had not been noticed is taken tentatively.

Experimental

The experimental procedure began with preparation of about 5 ml. of the test solution (C_{24} n-alkane on the order of 0.1 wt. % in heptane). Next, the solution was placed in a small glass vial of which the lower 1/3 was insulated by inserting it into a styrofoam cube. The insulated vial was then placed in a cylindrical channel through the wall of the Missimer FT4 cold chamber, which leads from the internal chamber to an external plexiglas window. The vial was placed as close to the window as the styrofoam cube would allow and was tilted to one side (see Fig. 2). Tilting the vial to one side and insulating the bottom 1/3 created a temperature gradient colder near the top of the vial. In this way crystals would grow from the top down and facilitate their observation and photography.

The initial temperature of the cold chamber was about -40°C. Once the solution vial was in place, the temperature set point was lowered in a step change to T_{set}=-70.6°C. About 20 minutes later the sample temperature was around -47°C and the cooling rate 0.35 to 0.45°C/min. Under these conditions (considerably super cooled) crystals began to form and were photographed at 10 second intervals. The t=0 in Fig. 1 is the time when the first photograph of the growth sequence was taken. That t=0 time is some time after nucleation. The cooling rate is believed small compared to the 10 second intervals used, so that quasi steady state conditions are believed to prevail.

The crystals were photographed with a 35mm Nikon F-2 camera using Kodak ASA 400 Tri-X film. The camera was equipped with motor drive, a 55mm lens with macro focus, and a polarizing filter. It was positioned by using a tripod and focussed with respect to the edge of the vial. The f-stop for the lighting conditions was f11.5. The lighting was a critical variable. For the crystals to be clearly visible, the light from a professional photographer's lens had to be reduced to a small beam incident to the tilted vial at an angle which provided maximum reflection tents, but not the surrounding area. Upon completion of a run, the vial was warmed to room temperature and the process repeated for several trials. To be able to accurately measure the relative crystal sizes, the photographs were enlarged about 10 times and printed at 5" X 7" size. The actual sizes of the crystals were determined by using the outside diameter of the vial as a reference dimension. The algebraic equation for the actual relative crystal length was

$$\frac{\text{Photographic crystal length}}{\text{Photographic vial diameter}} = \frac{\text{Actual relative crystal length}}{\text{Actual vial diameter (0.656" O.D.)}}$$

The actual crystal lengths were considered relative because the measurements were not from one end of the crystal to the other, but rather from a reference point on the photograph to one end of the growing crystal. Growth per time was the information sought, not actual size. Typical photographic data from this experimental procedure are shown in Fig. 3.

Summary

This note presents experimental data on the size of C_{24} n-alkane crystals, as they form in a dilute solution of chilled heptane.

The time intervals of observation are believed short compared to cooling rate. Thus, in the quasi steady state repeatable condition obtained, it is concluded that growth rates are constant for a given crystal but that growth rates are distributed over the individual crystals grown even if identical growing conditions prevail.

LITERATURE CITED

1. Abegg, C.F., J.D. Stevens, and M.A. Larson, "Crystal Size Distributions in Continuous Crystallizers when Growth Rate is Size Dependent," AIChE Journal, Volume 14, pp.118-22, 1968.

2. Bergland, K.A. "Formation and Growth of Contact Nuclei," Ph.D. Dissertation, Iowa State University of Science and Technology, Ames, 1981.

3. Branthaver, J.F., K.P. Thomas, S.M. Dorrance, R.A. Heppner, and M.J. Ryan, "An Investigation of Waxes Isolated from Heavy Oils Produced from Northwest Asphalt Ridge Tar Sands," Liquid Fuels Tech., Volume 1, pp.127-46, 1983.

4. Feldman, N. "Operability of Automotive Diesel Equipment at Temperatures Below Fuel Cloud Point," SAE 730677, June 18, 1973.

5. Holder, G.A. and J. Winkler, "Wax Crystallization from Distillate Fuels," J. Inst. Pet., Volume 51 pp. 228-52, 1965.

6. Holder, G.A., "Mechanism of Crystal Modification by Polymers," ACS Polymer Chem. Preprints 7 #1, pp.306-18, 1966.

7. McMillan, M.L. and S.R. Reddy, "Understanding the Effectiveness of Diesel Fuel Flow Improvers," SAE 811181, October 19, 1981.

8. Steere, D.E. and J.P. Marino, "Low Temperature Field Performance of Flow Improved Diesel Fuels," SAE 810024, February 23, 1981.

9. Turner, W.R., "Normal Alkanes," IEC Product Research Development, 10:238-60, 1971.

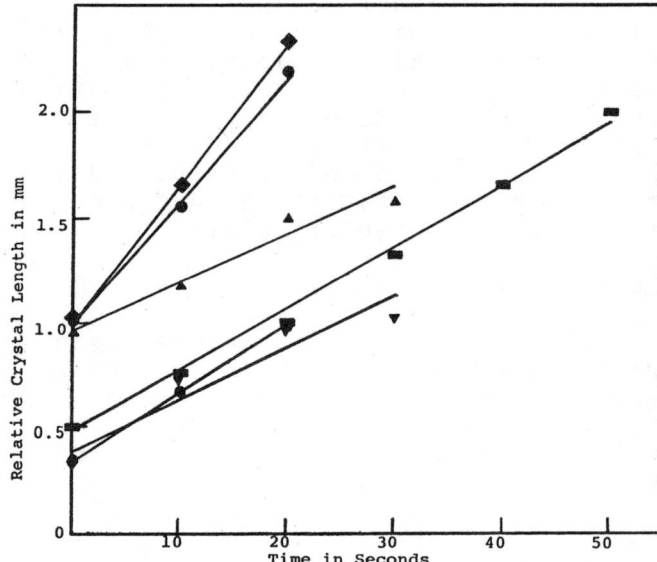

Figure 1. Growth rates of nC_{24} n-alkane crystals in a heptane solution show dispersion.

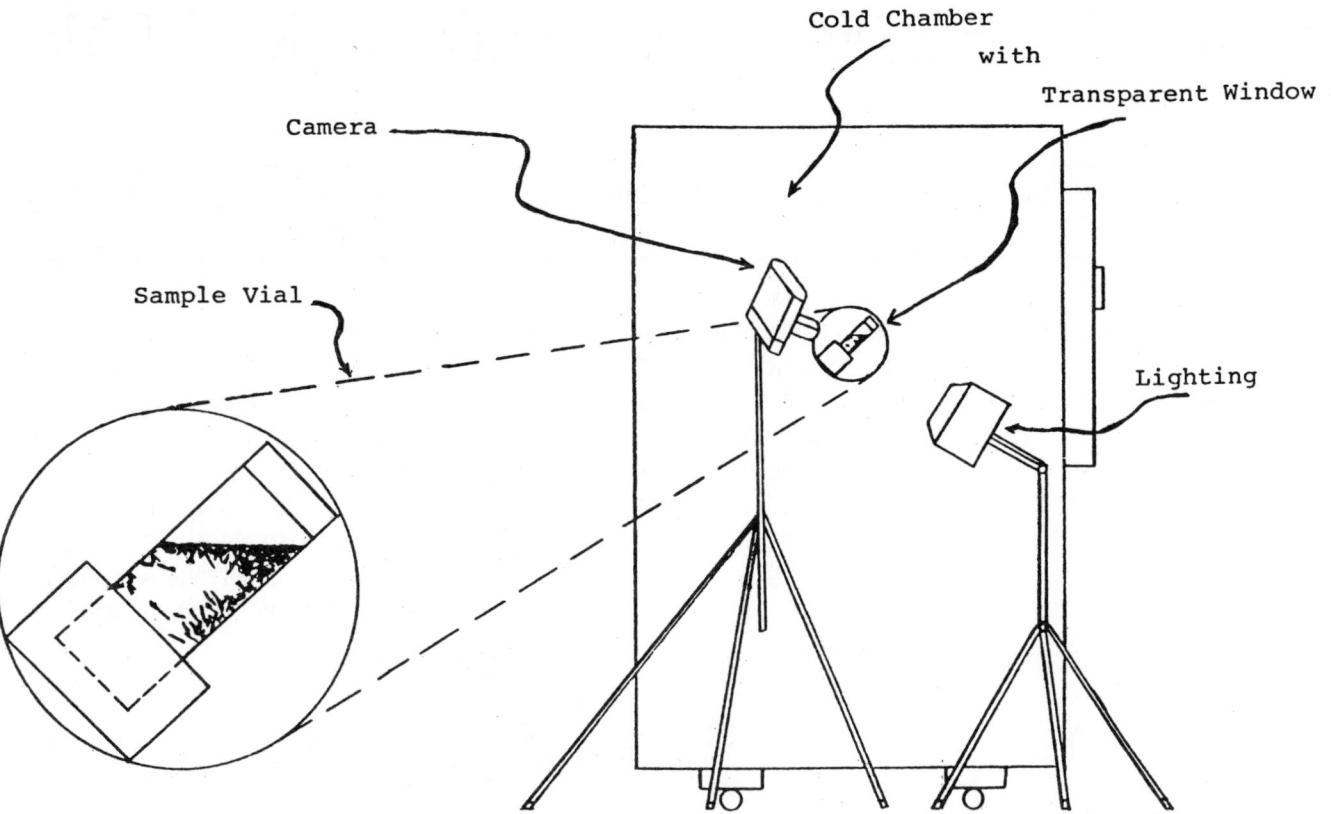

Figure 2. Elements of the experimental arrangement.

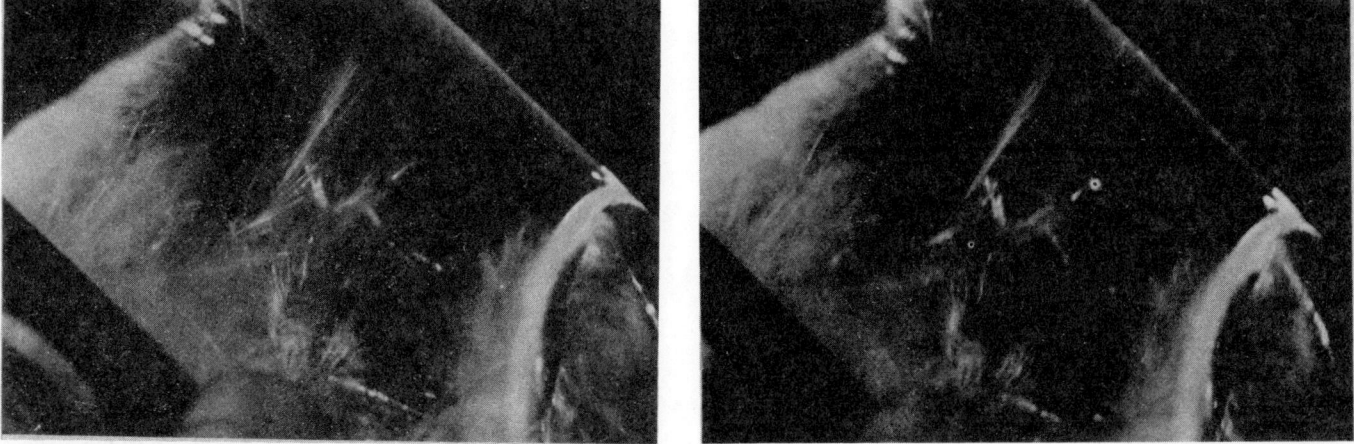

Figure 3. The bushy structure in the center of these photographs shows the relative growth of a C_{24} n-alkane crystal of two points in time.

THE GROWTH AND DISSOLUTION OF STRONTIUM FLUORIDE. INFLUENCE OF INHIBITORS.

A. Abdul-Rahman, S. M. Hamza, and G. H. Nancollas ■ State University of New York at Buffalo, Department of Chemistry, Acheson Hall, Buffalo, NY 14214

The influence of magnesium and zinc ions on the growth and dissolution of strontium fluoride crystals and of lead and aluminum ions on the dissolution of strontium fluoride has been studied in aqueous solutions at sustained super- and undersaturations using a constant composition method. The rates of crystallization both in the absence and presence of metal ions, expressed in terms of the relative supersaturation shows an effective order, n, of 4 ±0.1. The addition of magnesium and zinc ions even at relatively low concentrations reduced the rates of crystallization while the dissolution reactions were markedly inhibited by magnesium, lead, and aluminum ions. The influence of the metal ions on the strontium fluoride crystallization reactions could be interpreted in terms of a Langmuir-type adsorption isotherm. The kinetically derived adsorption affinities are compared with the directly measured adsorption equilibrium constants. Aging of the strontium fluoride crystal surfaces, which could be accelerated in the constant composition growth experiments, is an important parameter for determining the effectiveness of crystallization modifiers.

INTRODUCTION

The crystal growth of the alkaline earth fluorides is of importance in view of their applications in spectroscopy, electronics, lasers, glass manufacture, as well as their involvement in a number of industrial, biological, and environmental precipitation processes. The addition of fluoride ion to drinking water is now almost universal yet there is still considerable uncertainty about the manner in which it reduces the incidence of dental caries. Thus, Featherstone and co-workers (1) have reported that although carbonate ion is important in controlling the dissolution of tooth enamel, its activity can be markedly retarded when both strontium and fluoride ions are present. The factors that govern the mechanism of precipitation and dissolution of these fluoride salts are therefore of considerable interest, especially the influence of foreign cations which may exert a marked effect on the rates of crystallization and dissolution either through adsorption at the surface of the crystals or by lattice substitution. Despite the formation of these metal fluorides in the environment as a consequence of fluoride- containing industrial waste, the mechanisms of precipitation and dissolution are still in question especially in terms of changes in supersaturation, temperature, and ionic strength (2,3). In the present work, the constant composition (CC) potentiostatic method (4) has been used to investigate the influence of metal ions on the crystallization and dissolution kinetics of strontium fluoride. The solution compositions were maintained constant either by the addition of lattice ions, for the crystallization experiments, or of medium electrolyte for the dissolution reactions, using a fluoride ion-selective electrode in a potentiostatic mode.

EXPERIMENTAL

Solutions of strontium nitrate, potassium fluoride, potassium nitrate, magnesium nitrate, zinc nitrate, and lead nitrate were prepared using Ultrapure (Alfa Chemical Co.) chemicals with triply distilled, deionized water. The concentrations of divalent cations ($\pm0.2\%$) were determined by atomic absorption (Perkin Elmer Model 503) or by exchanging the metal for hydrogen ions on a Dowex-50 ion-exchange resin and titrating the liberated acid with standardized carbon dioxide-free potassium hydroxide (Dilutit, J.P. Baker Company). Solutions were prepared and stored in polyethylene or polypropylene vessels in order to prevent fluoride attack on glass surfaces. Strontium fluoride seed crystals were prepared as described previously (5) by precipitation using potassium fluoride and strontium nitrate solutions. They were washed repeatedly in triply distilled water and aged at the ionic strength of the experiments (0.072 mol dm^{-3} KNO$_3$) for periods ranging from one week to five months at 25°C. The pH of the seed slurries was maintained constant at 6.6$\pm$0.1 by the addition of dilute acid or

S.M. Hamza is now at Menoufia University Shebien El-Koum, Egypt.

base, and the solid phase was confirmed as strontium fluoride by x-ray powder diffraction (Phillips XRG 3000 x-ray diffractometer, Ni filter, and Cu-Kα radiation). Specific surface areas (SSA) of the seed crystals and of the grown phases were measured by BET nitrogen adsorption using a 10, 20 and 30% nitrogen/helium gas mixtures (Quantasorb II, Quantachrome, Greenvale, N.Y.). Scanning electron micrographs (ISI Model II, scanning electron microscope) showed the seed material to consist of small aggregates of cube-like crystals approximately 10 μm in size..

Growth and dissolution experiments were made in double-walled Pyrex 0.5 liter vessels with Teflon covers and polyethylene liners, in a stream of nitrogen gas presaturated at 25 °C with 0.072 mol dm^{-3} KNO_3 solution. The solutions were stirred using a suspended overhead magnetic stirrer in order to avoid grinding of crystals against the bottom of the vessel. Following the preparation of super- or under-saturated solutions, the reactions were initiated by inoculating either with a seed suspension of known slurry density, or with a weighed sample of dried seed. During crystallization experiments, the activities of the ionic species were maintained constant by the addition of two titrant solutions, consisting of (i) strontium and potassium nitrates and (ii) potassium fluoride. For the dissolution experiments, potassium nitrate background electrolyte was used as titrant. In each case, titrant additions were controlled using a fluoride specific ion electrode (Orion, Model 9409) and thermal electrolytic silver/silver chloride electrode together with a Metrohm combititrator (Model 3D Brinkmann Instrument Company). In order to avoid leakage of liquid junction solution into the cell, the reference electrode was separated from the cell solution by means of an intermediate potassium nitrate salt bridge of the same concentration as that in the cell. During experiments, aliquots were withdrawn from time to time, filtered (0.2 μm Millipore filters) and the solutions were analyzed for divalent cations by atomic absorption. The solid phases were investigated by scanning electron microscopy, x-ray diffraction, and specific surface area measurements.

Adsorption isotherms for both zinc and magnesium ions on strontium fluoride were obtained by equilibrating known weights of strontium fluoride seed crystals with the appropriate cation-containing solutions. Kinetic experiments indicated that adsorption equilibrium was attained within 6 min and 35 min for magnesium and zinc, respectively. In order to ensure equilibrium, however, equilibration times of at least 24 hr were used in the adsorption experiments.

RESULTS & DISCUSSION

The concentrations of free ionic species in the solutions were computed by successive approximations as described previously (5). The rates of crystallization, R_g and of dissolution, R_d, may be expressed in terms of the relative supersaturation, σ_g and undersaturations σ_d, respectively, by Equation 1:-

$$R_{g,d} = dn/dt = ks\, \sigma_{g,d}^{p} \qquad (1)$$

in which n is the number of moles precipitated in time t, k a rate constant, and s a function of the initial seed surface area. The relative super- and under-saturations are defined in terms of ionic products and solubility products for the metal fluoride salt as shown in Equation 2:-

$$\sigma_g = -\sigma_d = [(IP)^{1/3} - K_{so}^{1/3}]/K_{so}^{1/3} \qquad (2)$$

where the ionic activity products, IP, and solubility product, K_{so}, are expressed in terms of the appropriate activities of the ionic species, $[(a_{Sr^{2+}})(a_{F^-})^2]^{1/3}$ at time t and at equilibrium, respectively.

Previous kinetic studies of the crystallization and dissolution of strontium and magnesium fluorides indicate reaction orders, p, of 4 for strontium fluoride, (6) and 5 for magnesium fluoride (7) at higher supersaturations reflecting polynucleation processes. As equilibrium is approached, a parabolic rate law with p = 2 was found for both strontium and magnesium fluoride. Such similarities suggest that both salts grow by the same spiral growth mechanism at low supersaturation changing to a polynuclear crystallization at higher driving forces (6,8). For the corresponding dissolution reactions, the diffusion controlled dissolution at high driving forces also appears to change to a surface controlled parabolic rate law as equilibrium is approached. The demonstrated microscopic reversibility near equilibrium is especially interesting (5) since it is not possible to obtain reliable rate data under these conditions using conventional crystallization experiments in which the supersaturation is allowed to decrease.

Typical plots of titrant addition during strontium fluoride growth and dissolution are given in Figs. 1 and 2, respectively. It can be seen that the rates of both growth and dissolution are constant for more than 90% and 40% of reaction, respectively, suggesting that crystallization takes place on the added seed crystals without additional nucleation or spontaneous precipitation. This was confirmed by the scanning electron micrographs of the solid phases grown to an extent of more than twenty times the mass of inoculating seed. Although the crystals maintained their orthorhombic form, aggregation into larger particles occurred.

A typical plot of the specific surface area of the grown strontium fluoride phase as a function of the extent of reaction is shown in Fig.3. The decrease in SSA is more rapid than that calculated on the basis of an isometric three dimensional crystal growth. This reduction in SSA during the first 100% of crystal growth probably reflects crystal-lattice perfection and surface annealing processes. Indeed, the linearity of the growth curves in Fig.1 suggests that the decrease in SSA during this period over compensates for the general increase in surface area accompanying macroscopic crystal growth. Also included for comparison in Fig.3 is the specific surface area of strontium fluoride seed crystals aged in saturated solution for extended periods. Here again, it is seen that the general surface annealing process leads to a decrease in specific surface area similar to that attained during the growth of 60 mg of strontium fluoride corresponding to 100% of the initial inoculating seeds. Our constant composition studies have shown for both calcium phosphate and calcium oxalate that the use of grown material as seed in subsequent crystallization experiments results in a much greater retardation by added inhibitor molecules. The outgrowth of fast growing faces and accompanying reduction in the number of active growth sites during initial crystallization of aged seed crystals is especially interesting. The CC method enables large extents of growth to be achieved without additional nucleation at sustained supersaturations even at very low values of $\sigma_{g,d}$. Changes in the influence of inhibitor molecules during the growth of these phases provides new information concerning the pre-growth or ageing process.

The influence of added metal ions upon both growth and dissolution reactions are illustrated in Figs. 1 and 2, respectively. It can be seen that both magnesium and zinc ions markedly reduce the rates of crystallization and logarithmic plots of the rates of growth as a function of supersaturation in Fig.4 show that the effective orders of reaction with respect to supersaturation are the same both in the absence and presence of metal ions. The values of p = 4, again suggest polynucleation mechanisms for crystallization similar to that proposed in pure supersaturated solutions.

On the assumption that the decreased rate of growth in the presence of magnesium and zinc ions reflects their adsorption at active growth sites on the crystal surfaces, the degree of inhibition may be interpreted in terms of a simple Langmuir adsorption isotherm (9). This requires a linear relationship between the inverse of the relative reduction in rate, $R_0/(R_0-R_i)$ and the reciprocal of the inhibitor concentration. R_0 and R_i are the rates of crystallization in the absence and presence of metal ion, respectively. The applicability of the Langmuir model is demonstrated by the linearity of the plots in Fig.5. The adsorption affinity constants, given by the inverse slopes of the lines in Fig.5 are 16.4×10^4 dm^3 mol^{-1} for magnesium, and 1.7×10^4 dm^3 mol^{-1} for added zinc ions. It can be seen that the inhibition of crystallization is considerably greater in the presence of the highly hydrated magnesium ion. The results of typical adsorption experiments of magnesium and zinc by strontium fluoride surfaces are shown in Fig.6. The isotherms correspond to adsorption equilibrium constants of 13.7×10^4 dm^3 mol^{-1} for magnesium, and 1.5×10^4 dm^3 mol^{-1} for zinc ion. Although zinc ion markedly reduces the rate of crystallization of strontium fluoride, its influence upon the rate of dissolution, in contrast to magnesium, is much less as shown in Fig.2. The highly hydrated Al^{3+} shows a striking inhibition of dissolution, exceeding that by magnesium ion by a factor of more than five.

The influence of the crystallization driving force upon the degree of inhibition by metal ions is especially interesting. Plots of the kinetically derived adsorption affinity or equilibrium constant K_L, as a function of super- and under-saturations, shown in Fig.7, illustrate the striking increase near equilibrium. Experiments, made at low ionic strength in the presence of magnesium ion show a marked increase in the inhibition of dissolution suggesting a predominantly electrostatic interaction between the magnesium ion and strontium fluoride surfaces. Fig.7 illustrates the 10-12 times lower inhibition of strontium fluoride crystallization by zinc as compared with that of magnesium ion but with the same general increasing kinetic adsorption affinity constant with decreasing crystallization driving force. The adsorption affinity values calculated from the equilibrium experiments for both magnesium and zinc (corresponding to $\sigma = 0$) are shown in Fig.7. The marked dependence upon $\sigma_{g,d}$ of the effectiveness of growth and dissolution inhibitors has important consequences in assessing the usefulness of these compounds for industrial applications such as the control of scale. It is clearly insufficient to base such selections on the results of a limited number of threshold precipitation experiments. Moreover, where essentially electrostatic interaction is involved between the inhibitor and mineralizing surface, the ionic strength of the solution markedly influences the effectiveness of the scale control. The need for expressing the scale formation reactions in terms of the activities rather than concentrations of the lattice ions cannot be too strongly emphasized.

ACKNOWLEDGMENTS

We acknowledge financial support from the National Science Foundation in grant # CPE831338301.

LITERATURE CITED

1. Featherstone, J.D.B. Shields, C.P. Khademazad, B, and Oldershaw, M.D. J. Dent. Res., 62, 1049 (1983).

2. Shyu, L.S. and Nancollas, G.H. Croat. Chem. Acta., 53, 281 (1980).

3. Barone, J.P, Svrjcek, D, and Nancollas, G.H, J. Cryst. Growth 62, 27 (1983).

4. Koutsoukos, P, Amjad, Z, Tomson, M.B. and Nancollas, G.H, J. Am. Chem. Soc., 102, 1553 (1980).

5. Hamza, S.M, Abdul-Rahman, A. and Nancollas, G.H. J. Cryst. Growth, in press.

6. Bochner, R.A, Abdul-Rahman, A, and Nancollas, G.H. J. Chem. Soc., Faraday Trans. I, 80, 217 (1984).

7. Yoshikawa, Y. and Nancollas, G.H. J. Cryst. Growth, 69, 357 (1984).

8. Yoshikawa, Y. and Nancollas, G.H. J. Cryst. Growth, 64, 222 (1983).

9. Koutsoukos, P.G, Amjad, Z, and Nancollas, G.H. J. Colloid Interface Sci., 83, 599 (1981).

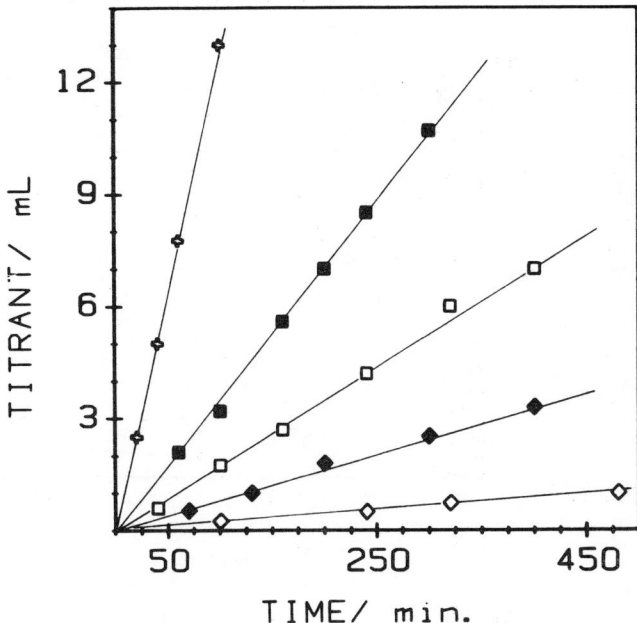

Figure 1. Strontium fluoride growth at $\sigma g=1.56$, molar strontium = ½ (molar fluoride) = 3.40×10^{-3} mol dm^{-3}, 25°C. Plots of SrF$_2$ titrant as a function of time ✢ no additives; ■, 9.50×10^{-5} mol dm^{-3} Zn(NO$_3$)$_2$; □, 2.00×10^{-4} mol dm^{-3} Zn(NO$_3$)$_2$; ◆, 2.40×10^{-5} mol dm^{-3}; Mg(NO$_3$)$_2$; ◇, 3.50×10^{-5} mol dm^{-3} Mg(NO$_3$)$_2$.

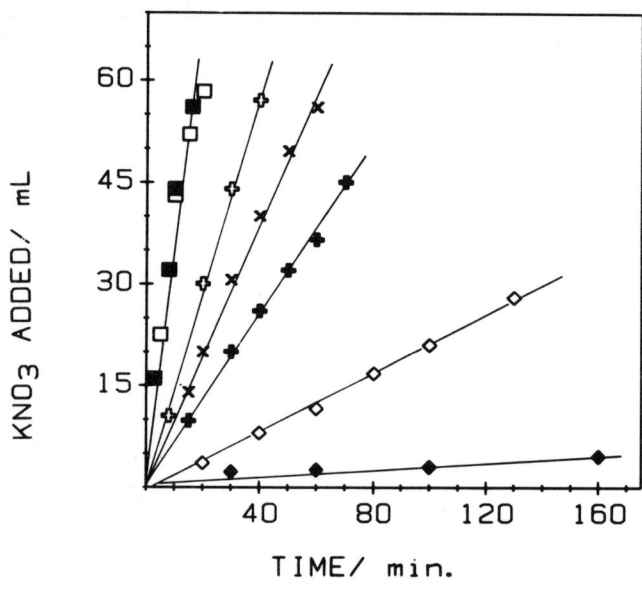

Figure 2. Strontium fluoride dissolution $\sigma_d = 0.40$, molar strontium = ½ (molar fluoride) = 0.80×10^{-3} mol dm^{-3}. Plot of medium electrolyte, KNO$_3$, as a function of time. ■, no additive; □, Zn(NO$_3$)$_2$ 5.0×10^{-4} mol dm^{-3}; ✢ Mg(NO$_3$)$_2$, 1.0×10^{-5} mol dm^{-3}; ✕, Pb(NO$_3$)$_2$ 1.0×10^{-5} mol dm^{-3}; ✚, Al(NO$_3$)$_3$ 1.0×10^{-6} mol dm^{-3}; ◇ Al(NO$_3$)$_3$ 2.0×10^{-6} mol dm^{-3}; ◆, Al(NO$_3$)$_3$, 5.0×10^{-6} mol dm^{-3}.

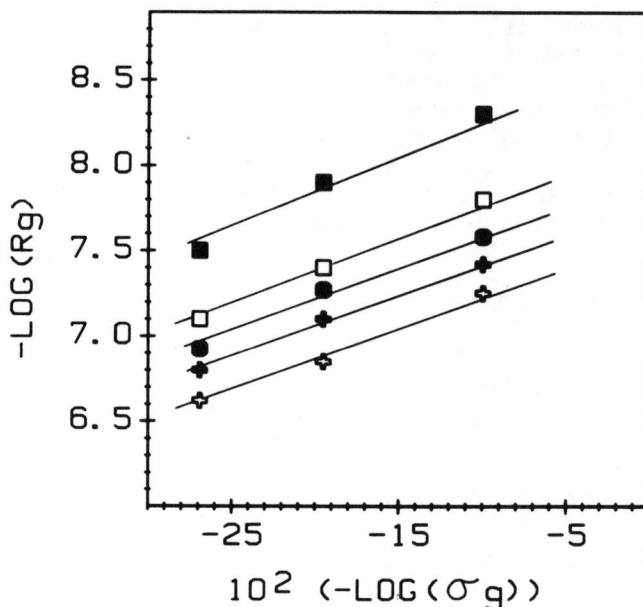

Figure 4. Logarithmic growth rate plots, R_g, against supersaturation, σ_g. Experiments in the absence of additives ✢; in the presence of zinc, ✚, 4.0×10^{-5} mol dm^{-3} and ●, 8.0×10^{-5} mol dm^{-3}, and in the presence of magnesium □, 1.0×10^{-5} mol dm^{-3}; ■ 2.0×10^{-5} mol dm^{-3}.

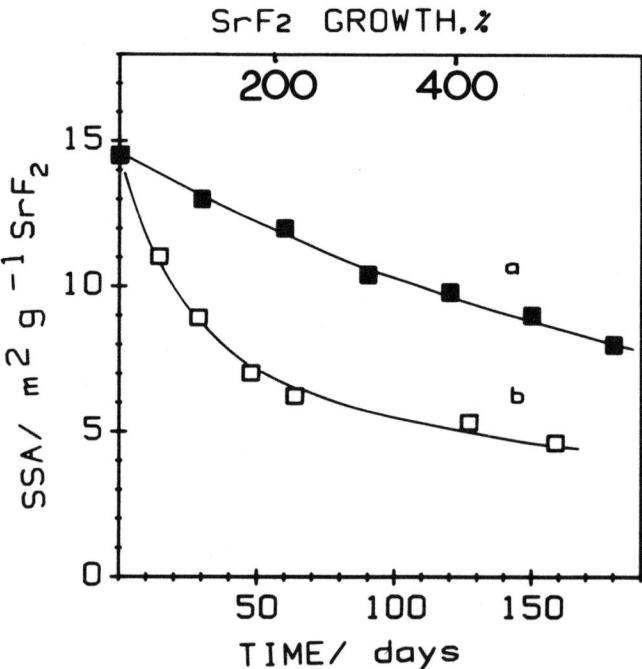

Figure 3. Plots of specific surface area as a function of (a) time of aging and (b) extent of growth.

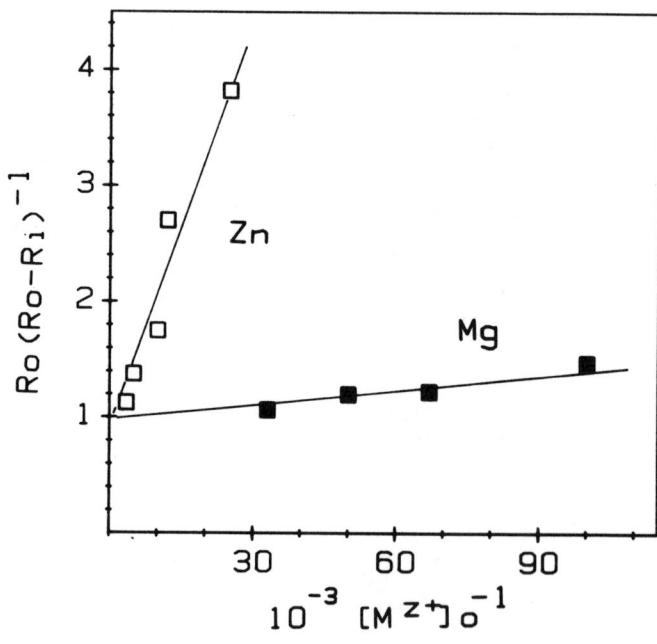

Figure 5. Langmuir plots of $R_o/(R_o-R_i)$ against $(M^{2+})^{-1}$ for the growth of strontium fluoride ($\sigma_g = 1.256$) in the presence of zinc, □, and magnesium, ■.

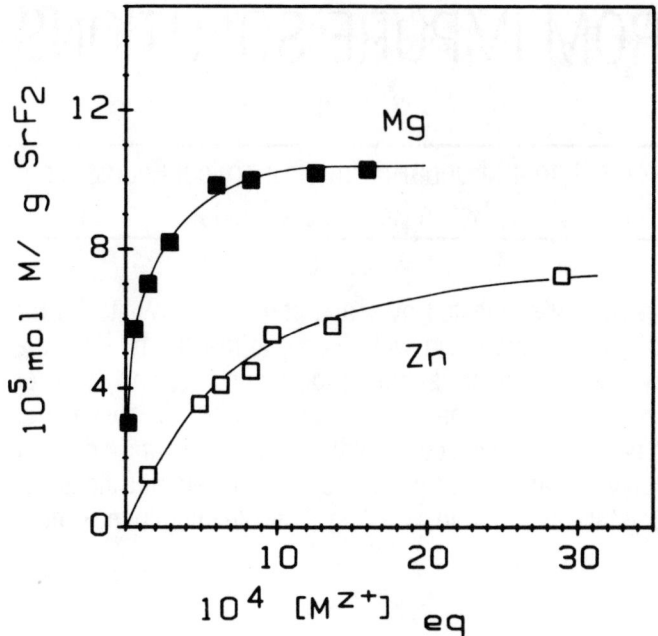

Figure 6. Adsorption isotherms of magnesium, ■, and zinc, □, on strontium fluoride at 25°C.

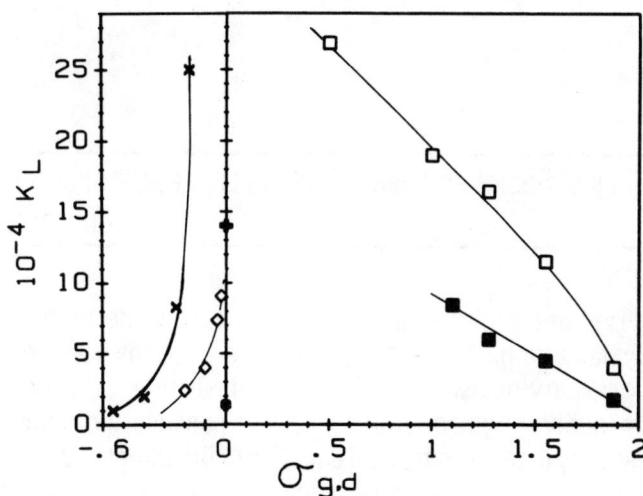

Figure 7. Kinetic adsorption affinity, K_L, as a function of growth (σ_g) and dissolution (σ_d) driving forces at ionic strengths of 0.072 mol dm^{-3} [growth, □ (magnesium), ■ (zinc with ordinate expanded 5X); dissolution, ✗ (magnesium)] and 0.15 mol dm^{-3} [dissolution, ◇ (magnesium)]. Equilibrium adsorption affinity, ✚, Mg; ●, Zn at $\sigma_{g,d} = 0$.

KINETICS OF SUCROSE CRYSTALLIZATION FROM IMPURE SOLUTIONS

Michael Saska and Jean-Paul Garandet ■ Audubon Sugar Institute and Department of Chemical Engineering, Louisiana State University, Baton Rouge, LA 70803

The empirical equation for crystallization rate of sucrose previously suggested by Maudarbocus and White (*1*) for crystallization of sucrose from impure solutions was not found in good agreement with our experimental data nor the preliminary measurements on the industrial crystallizers. A new, two-parameter equation for cooling crystallization of low-purity magmas is proposed that contains the mother-liquor viscosity as the independent variable. In absence of viscosity data, the empirical equation of Broadfoot and Steindl (*2*) was coupled with our growth-rate equation and a good fit was achieved with our experimental data. The work is continuing on verification of the growth rate equation for the industrial conditions. The equation was used in a computer model of the crystallizer to derive a cooling profile leading to maximum mother-liquor exhaustion within the constraints of the industrial installation.

Of the roughly 20 kg of sucrose lost in the manufacturing process for each 100 kg of sucrose in the input stream (sugarcane), 10 kg remain un-crystallized in the final mother-liquor. While the equilibrium limitations (i.e. sucrose solubility) account for much of the losses, the crystallization kinetics determines the size of the cooling crystallizers needed to approach equilibrium to a degree justifiable economically.

The goal of our work has been the determination of crystallization kinetics of sucrose at the conditions of industrial cooling crystallization, i.e. the temperature range 40-70°C, the sucrose/solute fraction in solution between 0.3 and 0.6 and solution viscosities of 10 to 100 Pa.sec and, eventually, development of guidelines for (1) parameters of the (inlet) magma (temperature, supersaturation, crystal content) and (2) the cooling regime of the crystallizers. A significant fraction of non-sucrose components in the crystallizing solution are of a high MW nature affecting primarily the solution viscosity, and also the crystallization rate and, in some instances, the habit of the sucrose crystals. In a concurrent work we have investigated the economic feasibility of removing the high MW substances by ultrafiltration with semi-permeable membranes of pore sizes 10^{-9} to 10^{-8} m. In order to provide a convenient expression for comparison of crystallization kinetics of sucrose from untreated and ultrafiltered solutions (that differ markedly in their viscosity but not as much in their solute content) it appeared desirable that the (empirical) crystal growth rate equation contain the solution viscosity as an explicit variable.

EXPERIMENTAL PROCEDURE

The industrial sucrose solutions were crystallized in a jacketed, horizontal crystallizer connected to a circulating water bath and a solid-state temperature programmer (Figure 1) at the stirring rate of approximately 1 rpm. The top of the crystallizer was sealed with a silicon sealant to prevent solvent evaporation. After seeding with sieve cuts of mean sizes between 2×10^{-4} and 4×10^{-4} m, the magma was cooled at a constant rate of about 1°C/hr from 70 to 30°C. Samples of the magma were taken periodically, crystals separated on a pressure-filtration funnel and the solution analyzed for total solids, sucrose and reducing sugars. Sucrose and reducing sugars contents were obtained by titration according to Lane-Eynon, the total solutes by refractometry on a 1:1 diluted solution. An empirical corrective equation (Matthesius and Mellet, *3*)

$$1/x_{TDS} = 1.013/x'_{TDS} + 0.00932/x_s$$

was used to correct for the systematic error of the refractometric determination.

The sucrose solubility for each crystallization experiment was obtained from the last sample, taken after the magma had been kept at the final temperature for about three days.

THEORY

The solubility ratio SC ($=C_{s,eq}/C^°_{s,eq}$) is assumed to be independent of temperature, and the solubility (mass fraction) of sucrose in water, $x^°_{s,eq}$, given by the polynomial

$$a_0 + a_1 T + a_2 T^2 + a_3 T^3 \quad (1)$$

where the coefficients a_0, a_1, a_2, a_3 were determined by Charles (4) to be respectively .644, 7.25×10^{-4}, 2.06×10^{-5} and 9.04×10^{-8}. In a well-mixed, batch, cooling crystallizer, the supersaturation σ (= $C_s/C_{s,eq} - 1$) is given at all times by the solution of the supersaturation balance

$$C_{s,eq} \, d\sigma/dt - s\sigma = s - G - N \quad (2)$$

With no solvent evaporation, the supersaturation rate s is given by

$$s = (-dT/dt)(dC_{s,eq}/dT) \quad (3)$$

where the solubility gradient is given by eq. 1 considering that

$$C_s = -(1 + C_{imp})/(1 - 1/x_s) \quad (4)$$

and

$$C_{s,eq} = -SC/(1 - 1/x^°_{s,eq}) \quad (5)$$

At the conditions of the experiments, the nucleation rate N was negligible. The few runs where nucleation was detected by observation under a microscope were not used in the evaluation of the growth kinetics. The overall mass growth rate (g/g magma x sec) is

$$G = 3\alpha n \rho l^2 R \quad (6)$$

where the shape factor $\alpha = .36$ and the characteristic length l was taken as the largest dimension of the crystal. An empirical equation proposed

$$R = P_1 (\sigma - P_2) \exp(-E_{act}/R' (1/(T + 273) - 1/333)) \exp(P_3 C_{imp}) \quad (7)$$

by Wright and White (5) for $\sigma > 1.5 P_2$ gives the linear growth rate of sucrose R as a function of supersaturation, temperature and composition. The equation was found to describe well the parallel-flow cooling crystallization of sucrose in industrial crystallizers equipped with rotating and reciprocating-type cooling elements (Maudarbocus and White, 1). Besides eq. 7 we considered a growth rate equation of the type

$$R = (Q_1 \eta^{Q_2})\sigma \quad (8)$$

where η is the mother-liquor viscosity (Pa.sec). An empirical correlation of Broadfoot and Steindl (2)

$$\eta = 0.11 (x_s/x_{TDS})^{-1.3} \gamma^{-0.16} \exp(3.7 (100 x_{TDS} - 0.19 (T-50))/(113.5 - (100 x_{TDS} - 0.19 (T-50))) \quad (9)$$

was found to apply to the industrial solutions used in our experiments for a wide range of x_{TDS} and was used in conjunction with eq. 8 to correlate the experimental results.

Both growth rate equations are illustrated in Figure 2 for four different compositions (constant for each curve). With increasing σ, R increases until reaching a maximum. Further increase in σ (i.e. lowering of T for the constant composition curves in Figure 2) results in progressively smaller R due to mass-transfer limitation in the liquid phase. The locus of the growth rate maxima is then the desirable (optimum) crystallization regime.

RESULTS AND DISCUSSION

An example of experimental data (run 4-17) is shown in Figure 3, where x_{TDS}, x_s and x_s/x_{TDS} are plotted vs. crystallization time. The initial period of dissolution is caused by the lack of a priori knowledge of the sucrose solubility and consequent difficulties in adjusting the initial supersaturation (by evaporation of water). Eq. 2 was integrated numerically with the initial conditions corresponding to the first experimental point on the descending part of the $x_s(t)$ curves. The parameters $Q_1 = 9.7 \times 10^{-7}$ m/sec and $Q_2 = -1.8$ of eq. 8 were found

that minimized the sum of differences-squared for all runs of the measured x_s values and those calculated from eq. 2. The eq. 2 integrated using both equations 7 and 8 is plotted in Figure 3. The equation of Wright and White (5) under-predicts, for this set of parameters, the crystallization velocity of sucrose.

Important is the problem of scaling-up of the results from our laboratory crystallizers to the industrial conditions. While the work is continuing in that direction, preliminary data from two Louisiana factories (Figure 4), equipped with 24m³ Blanchard-coil crystallizers connected in a series, indicate that eq. 8 over-predicts the rate of crystallization. This would not be unexpected since the mass-transfer conditions of the experimental crystallizer are likely to be more favorable than in the industrial crystallizers. Surprising is the degree to which eq. 7, tested by Maudarbocus and White (1) for industrial crystallizers of a similar type as those used in factories X and Y, over-predicts the crystallization rate of sucrose. It appears (Figure 5) that, compared to eq. 8, eq. 7 gives higher growth rates at low x_s/x_{TDS} (high x_{TDS}) and the ratio R_{ww}/R_{sg}

decreases with temperature (σ increases). In Figure 2 both equations are plotted for solution compositions corresponding to 0, 1, 2 and 3 hours of crystallization Y of Figure 4. As crystallization progresses, the optimum σ becomes smaller and the rate of crystallization within 3 hours drops to less than half of the initial value. The full "optimum cooling profile" (obtained using eq. 8) is plotted in Figure 6 together with the actual one, x_s/x_{TDS} and magma viscosity for

the crystallization Y of Figure 4. The magma should be heated by 4°C within an hour followed by cooling at an approximately 2°C/hr and progressively lower rate of cooling. The magma viscosity, calculated using eq. 9 and a correction on crystal content of Awang and White (6), reaches 480 Pa.sec at the end of crystallization.

NOTATION

a_0, a_1, a_2, a_3 — parameters of the solubility eq. 1

C — concentration; kg/kg water

E_{act} — activation energy of crystallization; J/kg

G — crystallization rate; kg/kg water sec

l — characteristic length of the crystal; m

n — number of crystals; 1/kg magma

N — nucleation rate; kg/kg water sec

P_1, P_2, P_3 — parameters of eq. 7 ($P_1 = 2.06 \times 10^{-6}$ m/sec, $P_2 = 0.005$, $P_3 = -1.75$)

Q_1, Q_2 — parameters of eq. 8 ($Q_1 = 9.7 \times 10^{-7}$ m/sec, $Q_2 = -1.8$)

R — the linear crystal growth rate; m/sec

R' — the gas constant; J/kg °C

s — supersaturation rate; kg/kg water sec

SC — solubility ratio ($C_{s,eq}/C^°_{s,eq}$)

t — time; sec

T — temperature; °C

x — concentration; kg/kg solution

x' — concentration measured by refractometry; kg/kg solution

Greek characters

α — the volume shape factor

ρ — crystal density; kg/m^3

η — solution viscosity; Pa.sec

σ — supersaturation $(C_s/C_{s,eq} - 1)$

γ — shear rate; 1/sec

Subscripts

eq — equilibrium

imp — impurity (non-sucrose, non-water)

s — sucrose

TDS — total dissolved solids

Superscripts

o — pure solvent $(C^o_{imp} = 0)$

LITERATURE CITED

1. Maudarbocus, S. M. R. and E. T. White, "Computer model of a cooling crystallizer", Proc. Qd. Soc. Sugar Cane Technol., 45, 45 (1978).

2. Broadfoot, R. and R. J. Steindl, "Solubility-crystallization characteristics of Queensland molasses", Proc. Int. Soc. Sugar Cane Technol., 17, 2557 (1980).

3. Matthesius, G. A. and P. Mellet, "An exhaustion formula for South African molasses", Proc. S. Afr. Sugar Technol. Assoc., 50, 206 (1976).

4. Charles, D. F., "Solubility of pure sucrose in water", Int. Sugar J., 62, 126 (1960).

5. Wright, P. G. and E. T. White, "A mathematical model of vacuum pan crystallization", Proc. Int. Soc. Sugar Cane Technol., 15, 1546 (1974).

6. Awang, M. and E. T. White, "Effects of crystal on the viscosity of massecuites", Proc. Qd. Soc. Sugar Cane Technol., 43, 263 (1976).

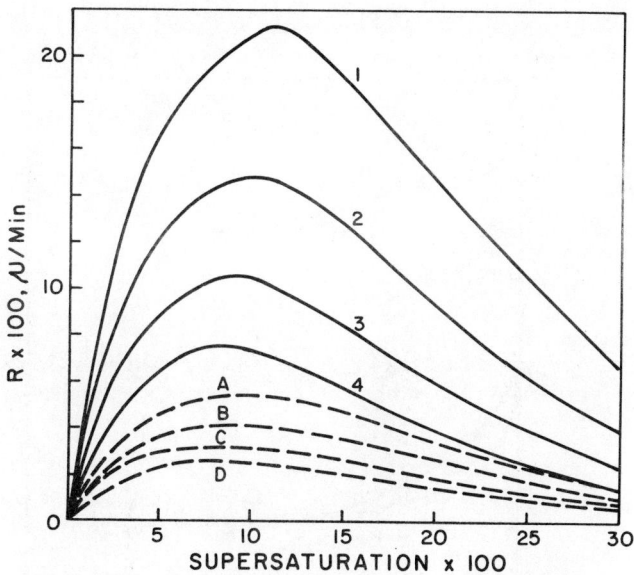

Figure 2. The crystal growth rate vs. supersaturation according to eq. 7 (1, 2, 3, 4) and 8 (A, B, C, D). The curves represent constant composition solutions corresponding to conditions of crystallization Y (Figure 4) at times 0 (curves 1 and A), 1 (2, B), 2 (3, C) and 3 (4, D) hours.

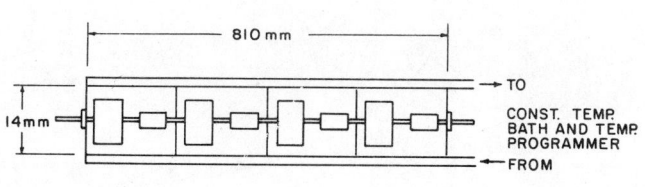
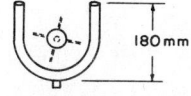

Figure 1. A laboratory cooling crystallizer.

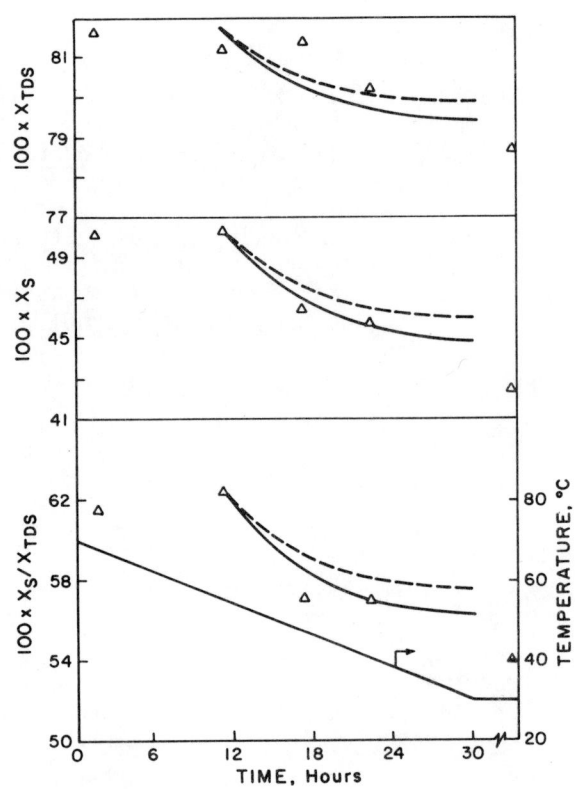

Figure 3. Results of experimental run 4-17 ( ). Curves represent the integrated eq. 2 using eq. 7 (- - -) and 8 (——) respectively.

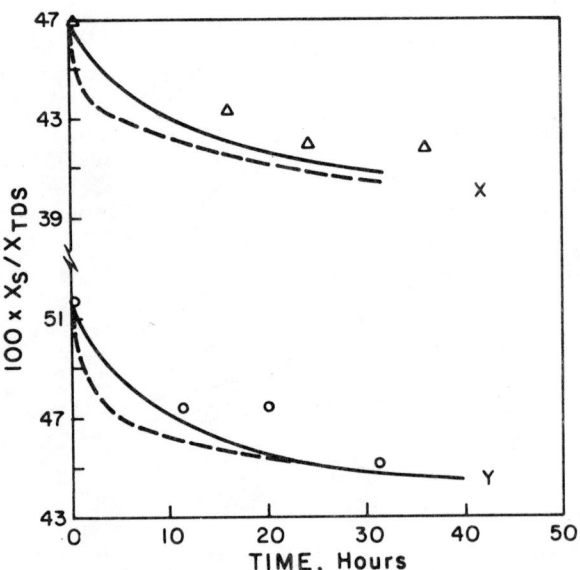

Figure 4. Integrated eq. 2 using the growth rate eqs. 7 (- - -) and 8 (——) for conditions of two industrial crystallizations X and Y.

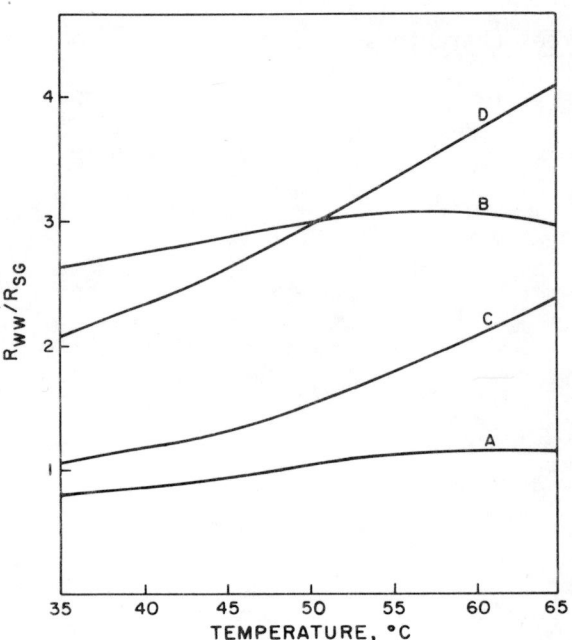

Figure 5. Ratio of crystal growth rates given by eq. 7 (WW) and 8 (SG) vs. temperature for constant composition solutions A, B, C and D.
A: x_S/x_{TDS} = 0.5, x_{TDS} = 0.833;
B: x_S/x_{TDS} = 0.5, x_{TDS} = 0.857;
C: x_S/x_{TDS} = 0.4, x_{TDS} = 0.862;
D: x_S/x_{TDS} = 0.4, x_{TDS} = 0.882.
The saturation temperature is approximately 67°C (A, C) and 80°C (B, D).

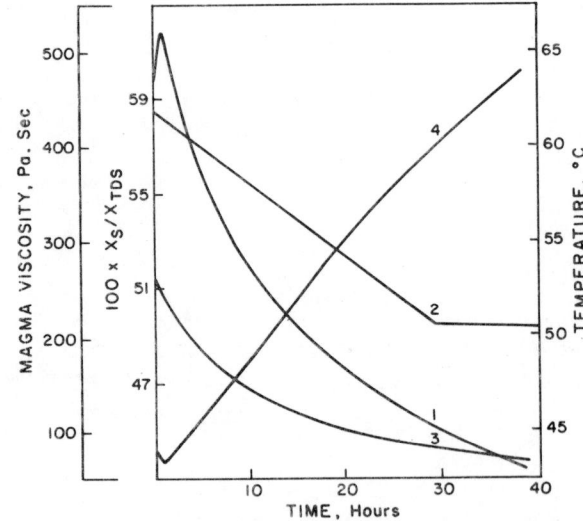

Figure 6. "Optimum" (curve 1) and actual (curve 2) crystallization temperatures at the conditions of the industrial crystallization Y of Figure 4. Curve 3: 100 x_S/x_{TDS}, Curve 4: magma viscosity, both at conditions of the "optimum" crystallization regime (curve 1).

SOLUTION GROWTH UNDER NONUNIFORM CONDITIONS— ADP AND MgSO$_4$ · 7H$_2$O

H. Narayanan ■ Clarkson University, Potsdam, NY 13676
G. R. Youngquist ■ Iowa State University, Ames, IA 50011

The breakdown of planar growth for single crystals of ammonium dihydrogen phosphate and magnesium sulfate heptahydrate has been studied using a submerged jet technique. Effects of temperature, jet Reynolds number and nonuniform supersaturation were interpreted using transport modeling combined with growth theory.

The characteristic shape of a crystal grown from solution often shows well defined planar faces. Since the growth of a crystal depends on the supersaturation at the solid-liquid interface, such planar growth often has been associated with uniform solute concentration at the interface. However, theoretical analyses (1,2) as well as experimental studies (3,4) of the related transport processes have shown that the interfacial solute concentration frequently is non-uniform even for cases where planar growth is observed.

Kumar, et al. (5,6) suggested that a submerged laminar jet of solution impinging on a growing crystal face, shown schematically in Figure 1, may provide some interesting insights. Due to growth of the crystal, the concentration (and hence the supersaturation) at the solid-liquid interface decreases radially from the stagnation point of the jet. Kumar solved the convective-diffusion equations for the solute concentration profile and coupled the results with crystal growth theory. This analysis suggested that a limited size region of uniform growth which is symmetrical about the jet will develop. Beyond this, breakdown of uniform growth should occur at a position where Wilson-Frenkel growth is attained. The analysis provided no information about the specific nature of the growth non-uniformity to be expected.

H. Narayanan is now with the Polaroid Corporation, Waltham, MA.

Subsequently, Kumar, et al. (6), and Narayanan, et al. (7) developed experimental methods which permitted observation of growth under conditions consistent with the analysis. Limited results for growth of magnesium sulfate heptahydrate and extensive data for growth of potassium aluminum sulfate from aqueous solutions confirmed the essential features of the analysis. A uniform or planar growth region developed opposite the jet, its size decreasing with increasing growth rate and decreasing jet Reynold's number. The breakdown of uniform growth was manifested in the formation of liquid inclusions and a stepped plateau structure outside the uniform growth region. In addition, by coupling the theory with experimental measurements, the growth coefficient $C = G/\sigma_{0min}$ was evaluated and reasonable agreement with the data of others was obtained.

In the present paper, results for the aqueous solution growth of ammonium dihydrogen phosphate (ADP) and additional data for magnesium sulfate heptahydrate are reported and provide additional experimental confirmation.

<u>EXPERIMENTAL</u>

The experimental apparatus is shown schematically in Figure 2 and is briefly explained here. The seed crystal M is placed with the face of interest perpendicular to the submerged jet C which provides a steady flow of solution. The solution flows past the crystal into the reservoir A from which the solution is recirculated continuously by the pump E. The entire apparatus is immersed in the water bath F for good temperature control. Seed crystals were grown from nucleates obtained by evaporation of solvent from a saturated solution at room temperature. These seeds were glued to a holder and allowed to grow to 1-2 cm. in mildly stirred, supersaturated solution. Saturated solution for the experiments was prepared by dissolving recrystallized salt in deionized water.

After washing with distilled water, a well-grown crystal was placed in the apparatus with the solution undersaturated by about 1°C. for at least an hour to avoid initial breeding. The solution saturation temperature was determined by the point of zero growth-dissolution by adjusting the bath temperature with the solution continuously circulated. Then the temperature of the solution was lowered to the desired undercooling. After attaining the desired supersaturation, the position of the crystal was adjusted with the aid of the collar K, and maintained at a distance of about 1 mm. from the tip of the jet. The growth rate of the region opposite the jet was measured by observing the linear advance of the face using a cathetometer. The experiment was designed to change the supersaturation by varying the growth temperature, keeping the solution concentration constant over a series of runs.

Qualitative features of the growing face could be seen with the naked eye and also were observed through the cathetometer. The experiment was continued long enough for the surface features to develop under steady conditions. After recording the growth rate, the crystal was removed and dipped in n-hexane to disperse the layer of solution adhering to the crystal surface. Photomicrographs then were taken to record the characteristics of the surface.

RESULTS AND DISCUSSION

The growth of ADP and magnesium sulfate heptahydrate was observed over a wide range of temperatures and supersaturations and a modest range of jet Reynold's number. Both systems behaved as anticipated, exhibiting uniform or planar growth opposite the jet, surrounded by inclusions or other types of non-uniform growth. There were significant differences in detail, however.

Qualitative Observations

For comparison, Figure 3 shows the growth behavior of potassium aluminum sulfate as previously reported by Narayanan, et al. (7). Under the influence of jet impingement, a plateau formed opposite the jet. The plateau had the same shape as the parent face, and had a circular central region where planar growth parallel to the parent face occurred. This central region was bounded by inclusions and ultimately by faceted steps, indicating the breakdown of uniform growth.

For ammonium dihydrogen phosphate, the growth temperature was varied from 2.0 to 33.2°C. with undercoolings from 1.8 to 18.0 °C. The jet Reynold's number ranged from 535 to 855. Figures 4 and 5 are front views of the (100) face of ADP crystals after impingement. As with alum, a circular uniform growth region developed opposite the jet. At moderate undercoolings, as in Figure 4, the uniform growth region was surrounded only by inclusions, which appear white in the photograph. At high undercoolings, as in Figure 5, the uniform growth region was surrounded first by inclusions. Beyond this, the breakdown of uniform growth gradually became more extreme. The uniform growth region was contained in an elevated plateau around which step structures could be observed.

At very low supersaturations, the breakdown of the uniform growth region was not apparent if the jet was located near the center of the crystal face. This was due to the fact that the size of the uniform growth region was larger than the face itself. However, if the jet was located at one edge of the face, the breakdown appeared near the other edge of the face in the form of a circle sector. When the undercooling was gradually increased during the run, the size of the uniform growth region decreased giving a sloped projection as in Figure 6. For very high undercoolings, the entire crystal face roughened. All of these observations are consistent with the theoretical analysis and also with the previous experimental results for potassium alum.

For magnesium sulfate heptahydrate, the growth temperature varied from 16.0 to 39.3°C. with undercoolings from 1 to 10°C. The jet Reynold's number varied from 115 to 200. The surface features responded as expected—uniform growth breakdown, formation of plateaus and liquid inclusions. These features were qualitatively consistent with the analysis. However, the details were quite different from those obtained with ADP and alum. The formation of plateaus and inclusions was dependent not only on surface supersaturation, but also on the growth temperature. At low growth temperatures, the uniform growth region was large, circular, and bounded by inclusions as in Figure 7 and similar to ADP and alum. The inclusions were thin tubular structures always oriented parallel to the long axis of the crystal. The lighting used in Figure 7 revealed that the uniform growth region was not optically flat, but is slightly pyramidal with the apex near the jet stagnation point. This phenomenon also was observed for alum, but not for ADP.

At high growth temperatures, the final results for magnesium sulfate were dramatically different from those a low temperatures. As the surface features first developed, a circular uniform region could be observed through the cathetometer. However, a faceted shape quickly evolved as is shown in Figure 8. This structure was prominently raised from the parent face and showed little evidence of inclusions.

Quantitative Analysis

The surface features such as the uniform growth breakdown and the formation of plateaus and inclusions responded more or less as predicted by Kumar's theoretical analysis. The linear growth rate G and the size of the uniform growth region were measured for each run. This experimental information can be used to calculate the growth coefficient C. The details of the underlying theory are explained by Kumar et al. (5,6) and summarized further by Narayanan, et al. (7). The theory shows the dependence of the surface concentration on radial position to be a complex function of the groups

$$\bar{w}_c = w_c/w_\infty \text{ and } N = N_G/N_{Re}^{3/4} N_{Sc}^{1/3}$$

where

$$N_G = G\rho_c r_0/\rho_s D, \quad N_{Re} = 2V^s r_0/\nu, \quad N_{Sc} = \nu/D.$$

Using the BCF theory, the growth rate can be represented by

$$G = C\sigma_0\{\tanh(\lambda_0/2\lambda_s)/(\lambda_0/2\lambda_s)\}$$

In order to maintain uniform growth with surface supersaturation decreasing from the stagnation point, the interstep distance λ_0 should decrease until the hyperbolic function reaches its maximum value of unity as λ_0 approaches zero, at which point the limiting Wilson-Frenkel growth rate is reached. For this limiting growth rate,

$$G = C\sigma_{0min} \quad (1)$$

where σ_{0min} is the limiting supersaturation attained at the radial location of the breakdown of uniform growth, r_{min}. The linear growth rate G of the uniform growth region was measured using the cathetometer. σ_{0min} cannot be observed directly, but may be computed at r_{min} using the transport model. This requires experimental measurement of the size of the uniform growth region, the bulk concentration, and the solution flow rate in addition to the growth rate. Whenever the uniform growth region was circular, the boundary was to be taken to be the onset of inclusions and measured with the cathetometer. When the region was faceted and non-circular, the area was measured and an equivalent diameter calculated for use in the computations. Bulk concentrations were evaluated from solubility data and saturation temperatures and also verified experimentally. Transport and physical properties were evaluated from literature data. Computational details are given elsewhere (8).

Using the measured G and the computed σ_{0min}, the growth coefficient C may be evaluated using Equation (1). C is expected to have Arrhenius temperature dependence given by $C = C_0\exp(-E/RT)$ where E is an apparent activation energy for growth. Figures 9 and 10 are Arrhenius plots of the data obtained for ADP and $MgSO_4 \cdot 7H_2O$, respectively. The open circles represent the results of the experiments. The solid lines in each case represent the least square fit of the data. Apparent activation energies of 6.6 ± 0.8 kcal/mole for ADP and 17.7 ± 1.2 kcal/mole for magnesium sulfate were obtained.

For ADP, there are few data available for direct comparison to these results. Alexandru (9) determined an overall growth rate for ADP from weight measurements and calculated a value of 1×10^{-4} cm/sec. for the average growth coefficient at 56.5°C. using the BCF growth law. Mullin, et al. (10) correlated growth data for ADP using a nuclei on top of nuclei growth expression which can be represented by

$$G = A\sigma^{5/6}\exp(B/\sigma)$$

Bennema, et al. (11) compared this expression with that of the BCF expression and estimated that A probably will be about the same order of magnitude as C or one order of magnitude higher. They suggested that a reasonable relation will be A $\approx$ 6C, giving values for the growth coefficient between 25 and 40°C. of about 1×10^{-4} cm/sec. They cite an apparent activation energy of 5.7 kcal/mole, which compares well with the present work.

Liu, et al. (12) measured the growth rate of the (110) face of magnesium sulfate and fit the data to the BCF growth expression. Their results along with those obtained by Kumar, et al. (6) are shown in Figure 10 and show very good agreement with the present experiments.

The general agreement of the present data with that of previous workers is quite good considering the disparate nature of the independent experiments involved. The scatter in the Arrhenius plots is of some concern. However, similar scatter often has been noted for crystal growth rates. On occasion this has been described as growth dispersion and attributed to variations in dislocation densities.

CONCLUSIONS

The growth of ADP and $MgSO_4 \cdot 7H_2O$ were examined using an impinging jet technique. Under the experimental conditions used, the variation in supersaturation along the crystal face resulted in the eventual breakdown of uniform growth. A uniform growth region opposite the jet, bounded by inclusions and plateaus, was obtained. The observations were qualitatively consistent with the theoretical analysis of Kumar (5) and with the previous experimental results of Kumar et al. (6) for $MgSO_4 \cdot 7H_2O$ and Narayanan et al. (7) for potassium alum. Growth coefficients were evaluated from transport modelling and growth data, using the boundary of the inclusion-free region to indicate the onset of the limiting growth. In terms of both magnitude and temperature dependence, agreement with the data of others was quite good.

NOTATION

C	Growth Coefficient (cm/sec)
D	Diffusivity (cm^2/sec)
E	Activation Energy (Kcal/mole)
G	Linear Growth Rate (cm/sec)
N	Dimensionless expression, $N_G / N_{Re}^{3/4} N_{Sc}^{1/3}$
N_G	Growth Number, $G\rho_c r_0 / \rho_s D$
N_{Re}	Reynolds Number, $2V^s r_0 / \nu$
N_{Sc}	Schmidt Number, ν/D
r_{min}	Dimensionless radial location at which breakdown occurs
r_0	Radius of jet (cm)
T	Temperature, K
V^s	Centerline velocity at the jet exit, (cm/sec)
w_c	Mass fraction of solute in the crystal
$\overline{w_c}$	Mass fraction ratio, w_c/w_∞
w_∞	Mass fraction of solute at the jet
λ_0	Interstep distance
λ_s	Mean distance for surface diffusion
ν	Kinematic viscosity, (cm^2/s)
ρ_c	Crystal density, (gm/cc)
ρ_s	Solution density, (gm/cc)
σ_0	Surface supersaturation
σ_{0min}	Surface supersaturation at which limiting growth occurs

LITERATURE CITED

1. Kuroda, T., T. Irisawa, and A. Gokawa, J. Crystal Growth 42, 41 (1977).

2. Wilcox, W. R., J. Crystal Growth 37, 229 (1977).

3. Berg, W. F., Proc. R. Soc. Ser. A 164, 79 (1938).

4. Bunn, C. W., Discuss. Faraday Soc. 5, 132 (1949).

5. Kumar, C., and J. Estrin, J. Crystal Growth 51, 323 (1981).

6. Kumar, C., J. Estrin, and G. R. Youngquist, J. Crystal Growth 53, 176 (1981).

7. Narayanan, H., G. R. Youngquist, and J. Estrin, J. Colloid Int. Sci. 85, 319 (1982).

8. Narayanan, H., PhD Thesis, Clarkson University, Potsdam, NY (1986).

9. Alexandru, H. V., J. Crystal Growth 10, 151 (1971).

10. Mullin, J. W., and A. Amatavivadhana, J. Appl. Chem. 17, 151 (1967).

11. Bennema, P., J. Boon, C. van Leeuwen, and G. H. Gilmer, Kristall and Technik 8, 659 (1973).

12. Liu, C. Y., H. S. Tsuei, and G. R. Youngquist, Chem. Eng. Prog. Symp. Ser. 67, No. 110, 43 (1971).

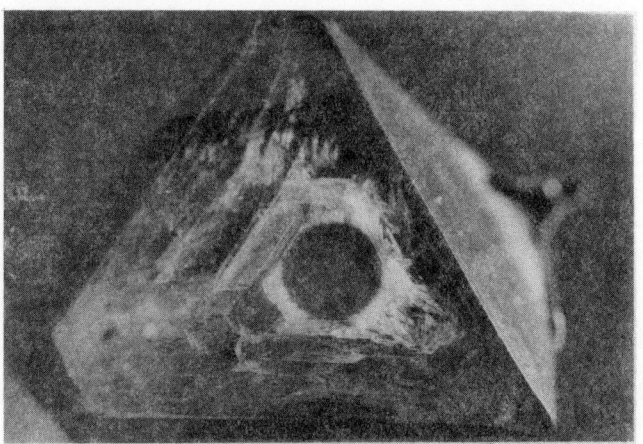

Figure 3. (111) Face of alum after exposure to the jet. (The circular central region exhibits uniform growth).

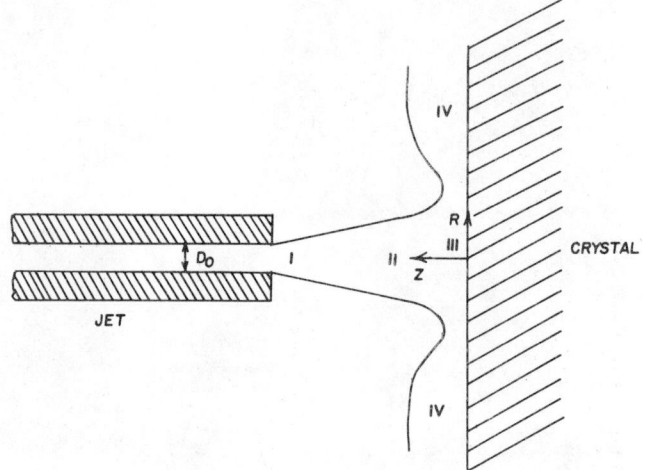

Figure 1. Schematic diagram of the impinging jet.

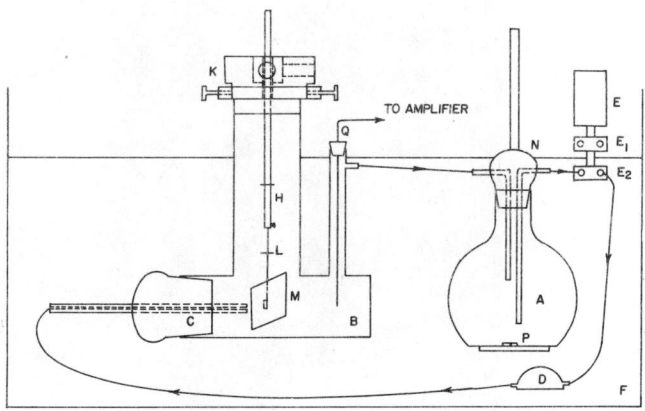

Figure 2. Schematic diagram of the apparatus:
A—solution reservoir; B—growth chamber; C—jet;
D—pressure equalizer; E—peristaltic pump; F—water bath
H—seed holder; K—adjustable collar; M—seed crystal
Q—thermocouple.

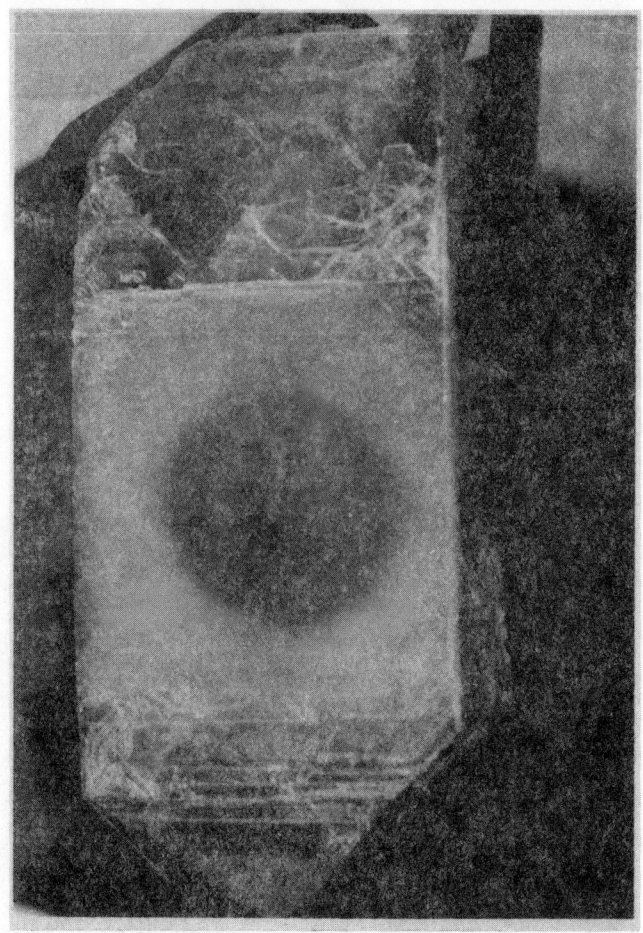

Figure 4. Face of ADP after exposure to the jet at low supersaturation.

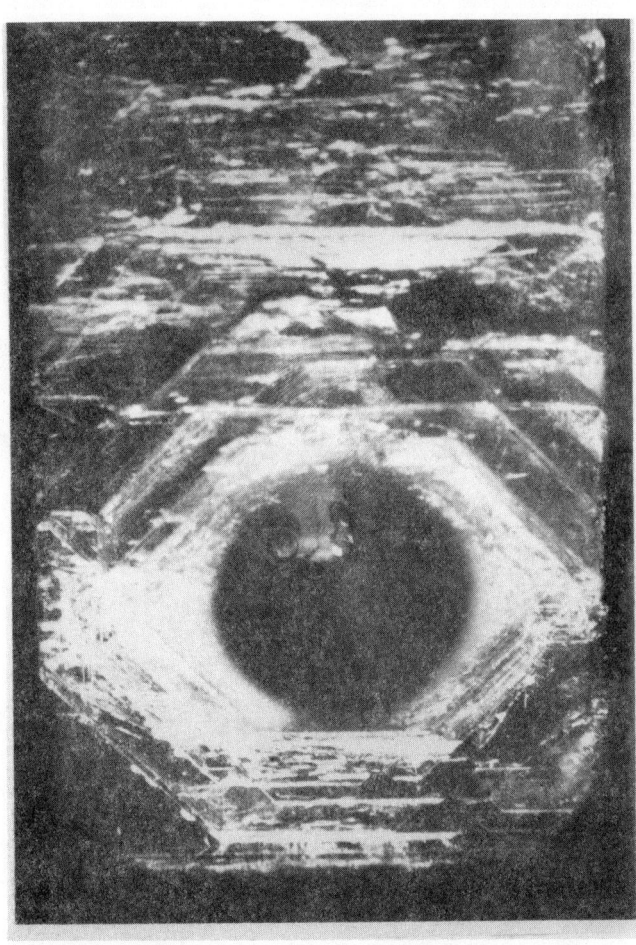

Figure 5. Face of ADP after exposure to the jet at high supersaturation.

Figure 6. Side view of the crystal plateau (ADP).

Figure 7. Central uniform growth region for $MgSO_4 \cdot 7H_2O$ at low temperature.

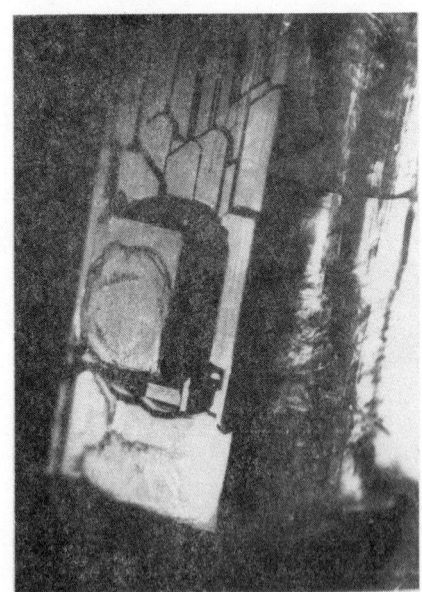

Figure 8. Growth on the (110) face of $MgSO_4 \cdot 7H_2O$ at high supersaturation and high growth temperature.

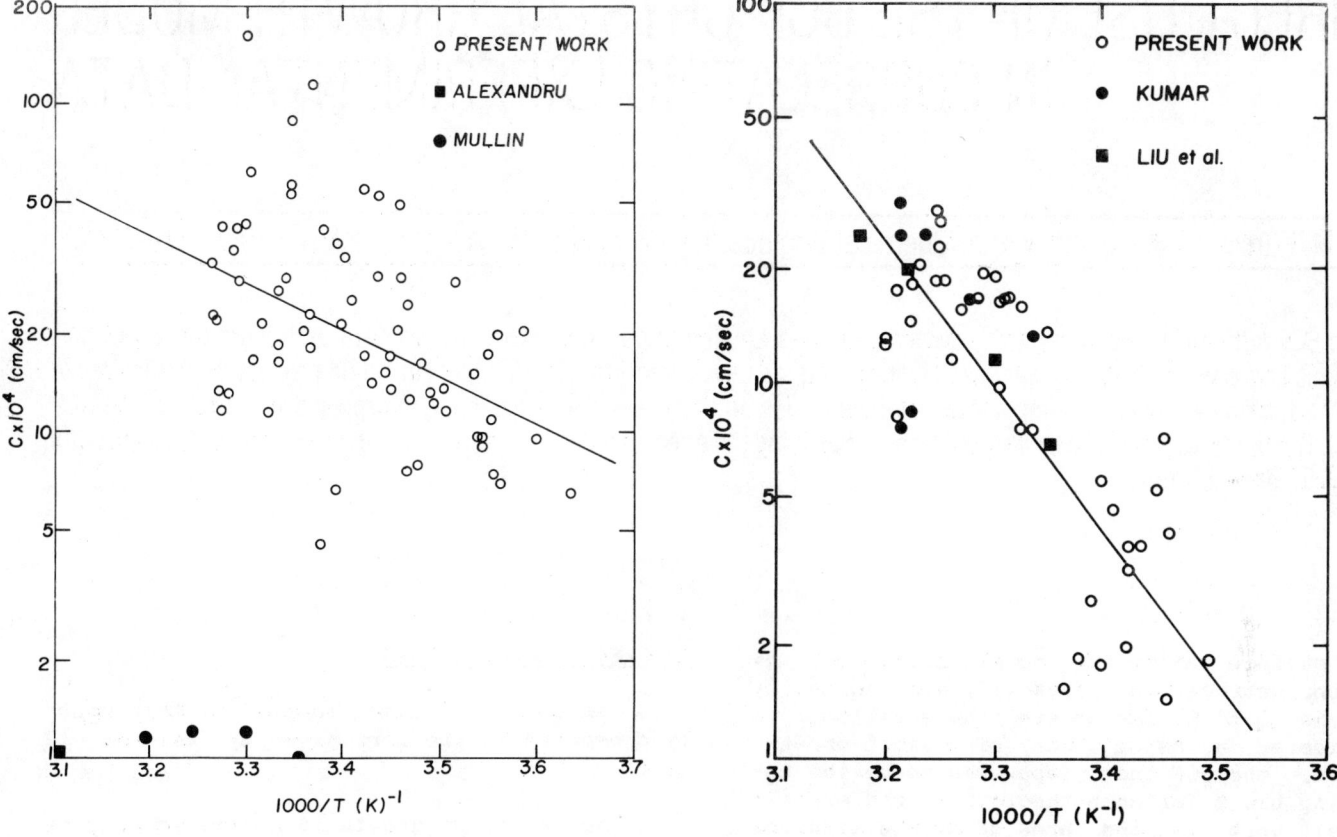

Figure 9. Dependence of the growth coefficient on temperature for ammonium dihydrogen phosphate.

Figure 10. Dependence of the growth coefficient on temperature for magnesium sulfate.

DIRECT USE OF THE BCF CRYSTAL GROWTH MODEL IN CORRELATING EXPERIMENTAL DATA

Piotr H. Karpinski ■ Department of Chemical Engineering, Worcester Polytechnic Institute, Worcester, MA

The BCF surface diffusion crystal growth model was applied to describe experimental data on the growth kinetics of inorganic salt crystals from an aqueous solution using the Levenberg-Marquardt nonlinear fitting procedure. The influence of hydrodynamic conditions, temperature, and impurities on the growth rates has been interpreted in the light of the BCF model. The model appeared to be very efficient in clarifying the effect of both the temperature and the presence of impurities on crystal growth rates.

The surface diffusion model introduced by Barton, Cabrera, and Frank (1), and adopted by Bennema (2 to 4) for growth from solutions, is considered a major theory of crystal growth. Indeed, the BCF theory appears to be the inspiring basis for both theoretical and experimental work in the area of crystallization [E.g., (2 to 11)]. Over the past three decades, many publications have been devoted to this theory. Although theorists consider it too simplified, its relatively complex mathematical form and a number of difficult to determine constants have been and still are major obstacles to its use. Many authors have tried with varying success to employ the BCF theory in describing the effect of external factors on crystal growth rates [E.g., (2 to 4, 7 to 10)], but the general uncertainties in the input kinetic data prevented the re-sults achieved from being very convincing. It is astonishingly surprising how readily far-reaching conclusions have been drawn based on experimental material of rather questionable quality.

The objective of this paper is to examine the usefulness of the BCF model in describing and interpreting very dependable experimental data concerning crystal growth from aqueous solutions under various conditions of flow-rate, temperature, and purity of the crystallizing system.

Worcester Polytechnic Institute, Worcester, Massachusetts, P.H. Karpinski is now with Eastman Kodak Company, Rochester, New York

THEORETICAL BACKGROUND

The basic BCF model adopted in this paper is presented in the form given by Bennema (2 to 4).

The rate of growth of a face growing by the spiral mechanism as a series of equidistant steps ($x_s >> x_0$) is described as follows

$$\dot{L} = \{[(\xi D_s c_o n_{so}\Omega ERT)/(9.5 x_s \tau a)]\sigma \ln(1+\sigma)\}\tanh\{(9.5\tau a)/[ERT x_s \ln(1+\sigma)]\} \quad (1)$$

Typically, the constants or rather quantities, since most of them depend on the supersaturation and temperature, of the above equation are grouped in the following two "model constants":

$$C = \xi D_s c_o n_{so} \Omega/(x_s)^2 \quad (2)$$

and

$$\sigma_1 = (9.5\tau a)/(ERT x_s) \quad (3)$$

and Equation (1) becomes

$$\dot{L} = (C/\sigma_1)\sigma \ln(1+\sigma) \tanh[\sigma_1/\ln(1+\sigma)]. \quad (4)$$

In addition, for most cases,

$$\ln(1+\sigma) = \sigma \quad (5)$$

and Equation (4) takes on the most celebrated form

$$\dot{L} = (C/\sigma_1)\sigma^2 \tanh(\sigma_1/\sigma) \quad (6)$$

Equation (6) can be subsequently simplified

for two extreme cases to yield

(a) the parabolic law if $\sigma_1 \gg \sigma$ (the supersaturation is small)

$$\dot{L} = (C/\sigma_1)\sigma^2, \qquad (7)$$

or (b) the linear law if $\sigma \gg \sigma_1$ (the supersaturation is large)

$$\dot{L} = C\sigma \qquad (8)$$

The physical situations and their mathematical pictures described by Equations (7) and (8) seem to justify a well-known power law form of the overall growth kinetic equation

$$\dot{L} = k_g \sigma^g \qquad (9)$$

where $1.0 \leq g \leq 2.0$. The interrelations among the quantities of Equation (1) are given below (2 to 4, 7):

The retardation factor

$$\bar{\Xi} = \{1 + (D_s t_s/ax_s) \tanh[\sigma_1/\ln(1+\sigma)]\}^{-1} \qquad (10)$$

The surface diffusivity

$$D_s \sim h^2 \exp(-E_D/RT) \qquad (11)$$

The mean distance diffused (the mean displacement)

$$x_s \sim h \exp[(E_d - E_D)/2RT] \qquad (12)$$

The growth spiral activity factor (8)

$$\epsilon = N/[1 + (y_o/2\pi r*)] \qquad (13)$$

DATA TREATMENT

Data used in the estimation of constants of the model have been collected over the past seven years by means of two basic kinds of experimental setups: fluidized bed crystallizers (12 to 15) and a flow cell (15, 16). The full documentation of the experimental conditions and the partial data interpretation focused on different scopes than those assumed in this work is described elsewhere (12 to 15, 17, 18). In addition, for comparative reasons, two sets of data of Hosoya et al. (19) were used.

All numerical computations and the data fitting, in particular, have been performed by means of the VAX-11/73 computer.

The Levenberg-Marquardt method of "nonlinear least squares" (20) was employed in assessing constants of the equations under consideration.

A relative mean square error defined as

$$s_i = 100 \cdot \{[(y_{calc} - y_{data})/y_{calc}]^2/n\}^{\frac{1}{2}} \qquad (14)$$

where the index i denotes a number of the respective equation, was chosen as a measure of the goodness of the fit.

The relative supersaturation, σ, occuring in the equations of the preceeding section, was calculated on the mass of hydrate basis; as defined below,

$$\sigma = (w - w*)/w* \qquad (15)$$

The errors, s_i, of the two-step growth kinetic equation (17, 18)

$$\sigma/\sqrt{(6\alpha\varrho_c \dot{L}/\beta)} = (1/k_d)\sqrt{(6\alpha\varrho_c \dot{L}/\beta)} + \sqrt{(1/k_r)} \qquad (16)$$

and of Equation (9) were used as reference errors.

RESULTS AND DISCUSSION

The constants of Equations (4), (9), and (16) were calculated and presented in Table 1 for five selected series of experiments. A typical dispersion of the experimental points is shown in Figure 1 for the series MgS.A. It is noticeable that all values representing the best fit were obtained in less than fifty computer program iterations.

The constants of Equation (6) appeared to differ from these obtained using Equation (4) by less than 5% and, thus, are not included in Table 1. This by no means surprising result is a simple consequence of the fact that the overwhelming majority of the supersaturation data was within the range 0.01 through 0.10. As shown by Ohara and Reid (7), the error of approximation resulting from Equation (5) is less than 5% for $\sigma \leq 0.10$.

The more striking fact, however, is that the simple power law's error was, as a rule, found to be smaller than the error obtained using the BCF equation in either form represented by Equations (4) or (6). This fact alone speaks against the unrestricted use of the BCF model. The maximum discrepancy was found when data concerning growth rates measured at different hydrodynamic conditions were compared. This discrepancy, in terms of the relative mean square error, reached as extreme values as 48% for Equation (4), versus 16% for Equation (9), for potassium alum, series ALUMB.D (u=2.3 cm/s, T=302 K). Consequently, the accuracy of the data fitting using Equation (4), was by far inferior to the accuracy obtained determining the standardized (18) constants of the two-step growth model, Equation (16). In fact, data describing the

Table 1. Constants of kinetic equations (4), (9), and (16) and associated correlation errors for selected experimental series.

Salt Studied	Principal Variable	$C \times 10^6$ m/s	$\sigma_1 \times 10^2$	s_4 %	$k_s \times 10^6$ m/s	g	s_9 %	$k_d \times 10^3$ kg/m²s	k_r kg/m²s	s_{16} %
K-alum T=302 K (13)	u=4.2cm/s	1.46	6.75	6.3	2.89	1.37	7.5	3.87	0.0617	3.2
	u=3.5cm/s	1.36	9.28	13	2.63	1.41	5.7	3.50	0.0437	2.4
	u=2.9cm/s	1.48	11.2	27	3.10	1.49	4.0	3.72	0.0410	2.7
	u=2.3cm/s	1.41	16.2	48	2.96	1.58	16	3.16	0.0303	7.3
	u=2.0cm/s	1.28	26.1	16	2.78	1.48	3.1	3.60	0.0341	2.0
	u=1.6cm/s	1.86	23.3	41	5.61	1.89	15	3.19	0.0216	10.4
K-alum u=2.9 cm/s (14)	T=286K	0.605	9.35	18	1.18	1.42	6.5	1.46	0.0216	4.2
	T=293K	0.864	8.61	14	1.71	1.41	7.0	2.31	0.0284	3.3
	T=301K	1.39	6.66	12	2.72	1.37	7.6	3.48	0.0633	3.6
	T=311K	2.19	6.67	12	4.45	1.38	7.3	5.74	0.0974	5.0
$MgSO_4 \cdot 7H_2O\text{-}Fe^{+3}$ T=301 K (15)	c_a=8 ppm	3.01	1.28	4.5	3.75	1.08	5.0	6.62	3.91	4.2
	c_a=19 ppm	2.44	2.20	7.9	3.80	1.18	4.5	5.83	1.11	4.1
	c_a=81 ppm	1.81	2.35	8.5	3.21	1.22	8.1	4.92	0.451	6.6
$MgSO_4 \cdot 7H_2O\text{-}Cr^{+3}$ T=301 K (15)	c_a= 0 ppm	1.93	2.27	5.7	4.85	1.32	6.0	5.73	0.420	4.4
	c_a=10 ppm	2.08	1.95	8.2	3.52	1.19	2.2	4.99	1.06	2.0
	c_a=27 ppm	2.05	0.734	3.7	2.41	1.06	3.9	4.13	11.6	3.6
	c_a=45 ppm	2.14	0	4.7	2.09	1.00	4.0	3.90	76.4	4.0
	c_a=165ppm	1.68	3.19	5.9	5.69	1.45	7.3	6.23	0.174	4.5
$Na_2S_2O_3 \cdot 5H_2O$ u=1.5 cm/s (14)	T=286K	9.71	0.916	4.2	11.1	1.05	4.6	13.4	21.6	4.2
	T=293K	11.4	2.90	27	26.3	1.32	7.1	24.2	1.15	5.6
	T=301K	11.2	1.86	5.8	16.1	1.14	5.5	17.6	4.47	4.9
	T=309K	13.2	0.803	5.2	15.9	1.07	3.5	17.9	41.4	3.3

effect of hydrodynamic conditions should not be interpreted via the BCF theory since one of its fundamental assumptions is that that the bulk diffusion process offers no resistance to the mass flux of a solute diffusing towards the step(s) of a growing crystal.

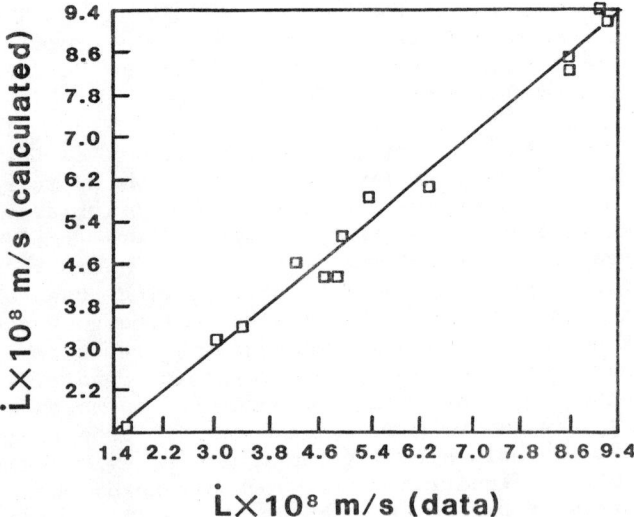

Figure 1. Calculated vs. experimental linear growth rate. Growth of $MgSO_4 \cdot 7H_2O$ in a fluidized bed crystallizer at T=301 K (12).

The errors of the best fit using the BCF surface diffusion model were still slightly larger than those of the power law and were always beyond the accuracy of the standardized two-step model, even when data obtained at constant hydrodynamic conditions were compared. The real power of the BCF model must, therefore, lie in a physical explanation of the particular behavior of experimental systems exhibited in studies of the temperature and impurity effect.

Effect of Temperature

A detailed analysis of Equations (2), (3), (10) to (13) suggests the <u>directions</u> of change of different quantities with the increase of temperature summarized in Table 2. Unfortunately, the magnitude of these changes is unknown. As a matter of fact, it is difficult to predict the resultant effect of the temperature in such a composite model as the BCF. Experience, however, clearly teaches that the growth rates are enhanced with the

Table 2. Directions of change of the constants of Equation (1) as the temperature increases.

Quantity	Expected Change	Other Conditions
Surface diffusivity, D_s	increase	
Mean diffusion distance on the surface, x_s	decrease	$E_d > E_s$
Saturation, c_e	probable increase	$E_s > E_d$
Retardation factor, $\tilde{s}$	slight increase	
Activity factor, ϵ	approaches a maximum	large σ

augmentation of the temperature.

The Arrhenius plot for the constant C is shown in Figure 2. The apparent activation energies were found to be $E_a = 3.9 \times 10^4$ J/mol for potassium alum and $E_a = 8.6 \times 10^3$ J/mol for sodium thiosulfate pentahydrate. More important than their magnitudes, indicating the overall energetic resistance to the growth process, is their ratio which corresponds to that found previously in studies on the overall growth kinetics involving these two salts (14). Clearly, the surface diffusion process in the case of potassium alum is energetically less favored than the same for sodium thiosulfate.

As shown in Figure 3, the constant σ_1 also depends on temperature. The analytical form of this relation can be deduced from Equations (3) and (12) to give

$$(1/\sigma_1) = AT \exp(B/T) \qquad (17)$$

where A and

$$B = (E_d - E_D)/2R \qquad (18)$$

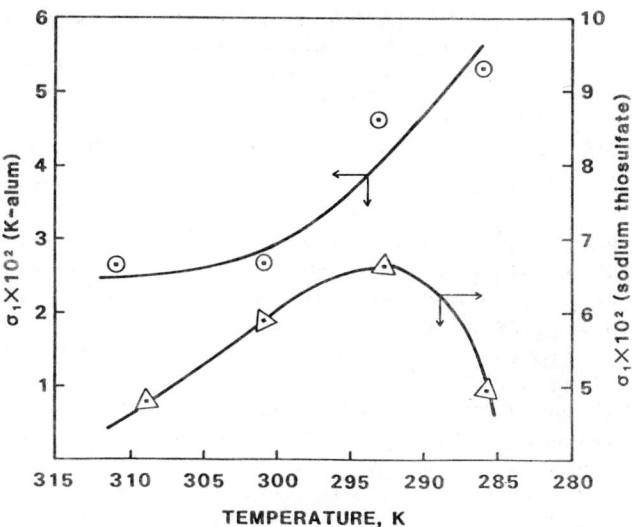

Figure 3. Dependence of the model constant σ_1, on temperature. The systems studied as in Table 1 and Figure 2, ref. (14).

are constants independent of temperature. The positive value of the constant B for a given system would, therefore, mean that $E_d > E_D$. The plot $\log(1/\sigma_1 T)$ vs. $1/T$ for potassium alum is shown in Figure 4. The value of the activation energy difference, defined by Equation (18), was found to be -1.8×10^4 J/mol. This value physically means that the desorption process is energetically favored over the surface diffusion, contradicting the somewhat arbitrary estimation of Bennema (4).

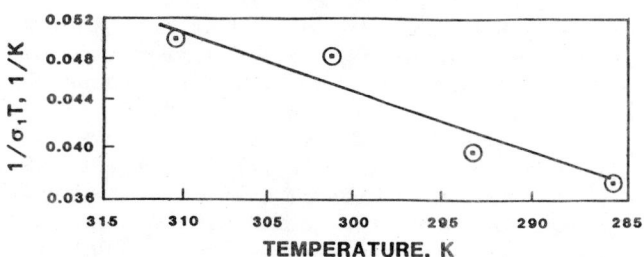

Figure 4. Illustration of use of Equation (17) for K-alum.

As can be seen from Figure 3, the curve describing the dependence of σ_1 on temperature for sodium thiosulfate exhibits a maximum. In fact, the presence of this maximum is mathematically predictable since the first and second derivatives of the reciprocal of the constant σ_1 are as follows

$$d(1/\sigma_1)/dT = A \exp(B/T) - (AB/T) \exp(B/T) \qquad (19)$$

and

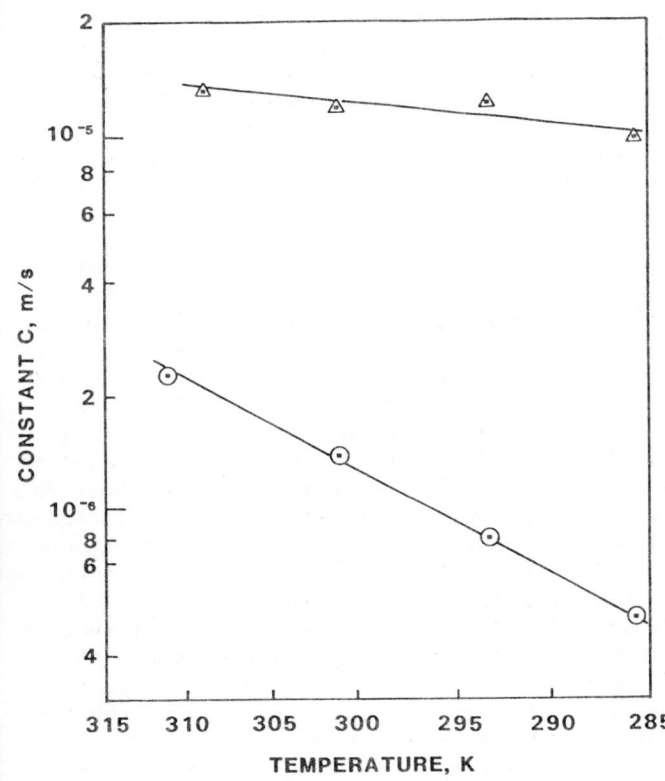

Figure 2. The Arrhenius plot for the model constant C. Growth in a fluidized bed crystallizer (14). ○ – K-alum, u=2.9 cm/s; △– Na$_2$S$_2$O$_3$·5H$_2$O, u=1.5 cm/s.

$$d^2(1/\sigma_1)/dT^2 = (AB/T^2)\exp(B/T) - (AB/T^2)\exp(B/T)$$
$$+ (AB^2/T^3)\exp(B/T)$$
$$= (AB^2/T^3)\exp(B/T) > 0 \quad (20)$$
$$(\text{a minimum!})$$

Letting $d(1/\sigma_1)/dT = 0$ gives $B = T$, or

$$E_d - E_s = 2RT \quad (21)$$

Thus, the proven existence of a minimum for the function $1/\sigma_1 = \emptyset(T)$ also proves that there is a maximum value for σ_1 itself. This fact is reflected in the data of Table 1 and of Figure 3. In other words, for systems having a positive difference between the activation energies characterizing the surface diffusion and the surface desorption of the growth units on the range of 5×10^3 J/mol, one should expect that a maximum will occur in the plot σ_1 vs. T, providing that the range of temperatures covered is sufficiently wide. For sodium thiosulfate, this maximum, $E_d - E_s = 4.9 \times 10^3$ J/mol, occurred at $T \approx 293$ K, cf. Figure 3. A positive value of the activation energies difference, as defined by Equation (21), indicates that the process of the surface diffusion is for this salt energetically preferred to the desorption process. This fact coincides with a well-known fact that the surface integration process for the sodium thiosulfate is very rapid (large k_r values; $g = 1.0$).

Effect of Impurities

The effect of impurities on different constants of Equation (1) is by far less clear than the effect of the temperature. Not only the magnitudes but often also the directions of possible changes of these constants are uncertain. Therefore, it is extremelly difficult to predict the resultant change in growth rates based on the BCF model alone. Again, the experimental results have to be utilized. All literature reports agree that for the concentration of impurities exceeding the ≈20 ppm level the growth rates are suppressed below those characterizing pure systems. The situation, however, may be opposite for the impurity concentration levels below ≈20 ppm as documented previously (15). In this author's opinion, an impurity has a twofold effect. First, the impurity lowers the interfacial potential and thus acts in favor of the surface incorporation of the principal solute. Second, it adsorbs on the active centers of the growing crystal's surface and thus hinders the growth units of the principal solute from their surface incorporation. The first effect prevails at the low impurity concentration level, when the blocking effect is not yet pronounced, whereas the second becomes quickly appreciable with the increasing impurity concentration level.

Figures 5 and 6 present the effect of impurities on both model constants.

As readily follows from the definition of σ_1, Equation (3), the lower the free edge energy of a growth unit in a step, τ, the lower the σ_1 value. Similarly, the lower the interfacial potential, the more growth units of the principal solute, n_{ao}, which may enter the adsorption layer. At low impurity concentration the impurity may decrease the surface energy (21, 22). This is true in particular for mobile or partially mobile impurities which, according to Cabrera and Coleman (5), lower the edge free energy, τ, by impurity adsorption in the step. These facts, reflected in the results of Table 1 and in Figures 5 and 6, explain the increase in C and diminution of σ_1 for low concentration of the impurities.

The blocking effect, predominant at higher impurity concentrations, may be explained in the following way. The number of the growth units of the principal solute which can be accomodated in the adsorption layer, n_{ao}, decreases and, possibly, the constant c_o diminishes. Similarly, since the relaxation time for the incorporation of a growth unit into the step, t_k, is prolonged due to the obstructive action of the impurity, Ξ, defined by Equation (10), will decrease. All these facts result in the decrease of the model constant C. The model constant σ_1 clearly tends to increase at the higher impurity concentration levels. Only two quantities, x_s and ϵ, can be considered to explain the course of the experimental curves in Figure 6. Mathematically, these constants should diminish in order to augment the σ_1. The diminution of x_s, however, would very strongly tend to increase the model constant C, which is proportional to $(x_s)^{-2}$, contradicting the experimental evidence. Thus, from the viewpoint of the mathematical formalism, the growth spiral activity factor, ϵ, is one that should become smaller to support the results of experiments. The adopted physical picture, according to which impurity entering the adsorption layer blocks the kinks, also speaks for diminution of ϵ. The blockage of the kinks causes an enlargement of the distance between steps in a parallel sequence and, thus, reduces the number of simultaneously developing spirals, N, and, consequently, makes ϵ to decrease in the conformity with Equation (13).

Finally, one more fact of some general importance seems to be worth of notice. The

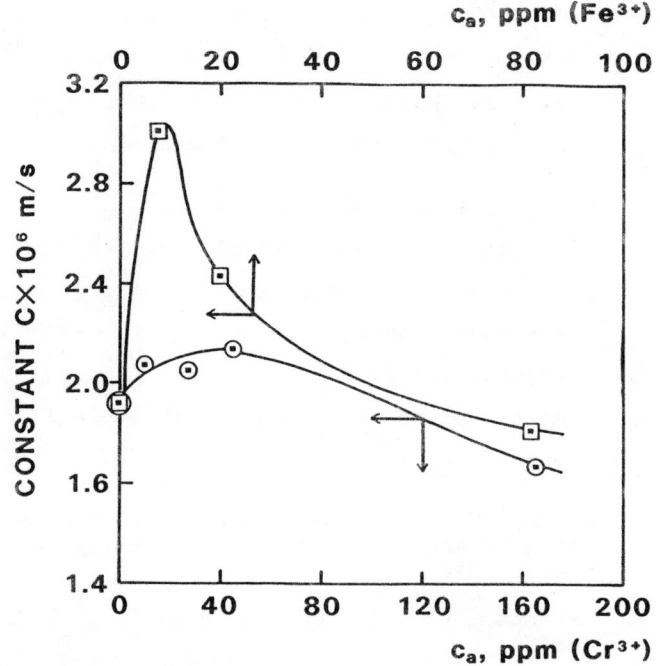

Figure 5. The effect of cationic impurities on the model constant C. Growth of $MgSO_4 \cdot 7H_2O$ in a fluidized bed crystallizer at T=301 K (15). □ – Fe^{+++}, ○ – Cr^{+++}

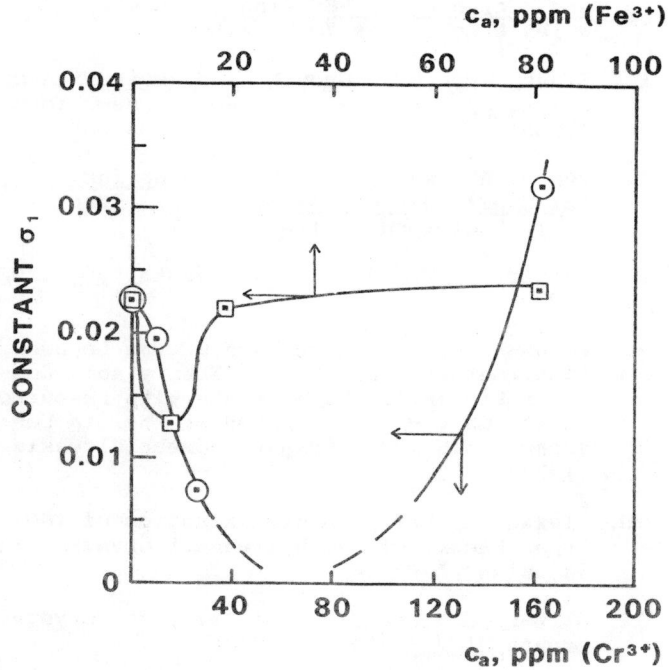

Figure 6. The effect of cationic impurities on the model constant σ_1. Growth of $MgSO_4 \cdot 7H_2O$ in a fluidized bed crystallizer at T=301 K (15). □ – Fe^{+++}, ○ – Cr^{+++}.

two-step growth model enables one to calculate the driving force for either of its steps. E.g., the driving force for the surface reaction, σ_r, can be calculated as

$$\sigma_r = (k_s \sigma^*/k_r)^a \quad (22)$$

and according to its definition (17, 18), this is the overall driving force minus the driving force for the diffusion step. A thoughtful application of the BCF model requires that the bulk diffusion contribution to the flux of solute transferred from the solution to crystal phase be eliminated. This elimination can be done by the use of a "bulk diffusion-free" driving force given by the above equation. Obviously, kinetic data treated in such a manner would give the statistically best fit, with the correlation error practicaly identical with that found for Equation (16), for

$$\tanh(\sigma_1/\ln(1+\sigma_r)) = 1.0 \quad (23)$$

Then the linear growth rate would become

$$\dot{L} = C^* (\sigma_r)^a \quad (24)$$

where $C^* = [\beta/6\alpha\varrho_c]k_r$, whereas C and σ_1 constants would become indistinguishable.

All these closing remarks are in accordance with the standardized constants of Equation (16) concept (18) whose application has demonstrated that for all the available growth kinetic data the assumption of the parabolic law for the surface integration is statistically meaningful.

CONCLUSIONS

The BCF surface diffusion model has appeared to be less accurate in the experimental data fitting than the two-step growth model and than the power law as well.

Due to its intrinsic limitations the BCF model should not be used for the explanation of the effect of hydrodynamic conditions on the growth rates.

The model has, however, been successfully applied to clarify the effect of the temperature and presence of impurities on both kinetics and mechanism of growth of inorganic salts from aqueous solution.

The method of assesing data concerning the interrelation between the activation energies of the surface diffusion and surface desorption processes has been introduced and applied to explain the dependence of the model constants, C and σ_1, on temperature.

It has been shown that the cationic impurities at the concentrations exceding ≈20 ppm may reduce the number of cooperating growth spirals.

NOTATION

a	shortest distance between neighbouring growth units in the crystal, m
A	constant in Equation (17), 1/K
B	constant in Equation (17), K
c_a	concentration of impurity, ppm
C	BCF model constant, m/s
D_s	surface diffusivity, m²/s
E_A	molar activation energy for the overall growth process, J/mol
E_d	molar activation energy for the surface desorption, J/mol
E_s	molar activation energy for the surface diffusion, J/mol
g	overall growth order
h	step height, m
k_d	mass transfer coefficient, kg/m²s
k_g	overall growth constant, m/s
k_r	surface reaction rate constant, kg/m²s
$\dot{L}$	linear growth rate of a face, m/s
n	number of experimental points
$n_{s,o}$	number of growth units in the adsorption layer
N	number of the cooperating growth spirals
r^*	radius of critical nucleus, m
R	universal gas constant, J/K mol
s_i	relative mean square error of Equation (i), %
t_k	relaxation time for entering the kink, s
T	temperature, K
u	superficial velocity of solution, m/s
w	actual mass concentration, hydrate basis kg/kg
w^*	saturation concentration, hydrate basis, kg/kg
x_s	the mean diffusion distance on the crystal surface, m
y_{calc}	calculated value of a quantity using the best fit correlation
y_{data}	measured value of a quantity
y_o	distance between steps, m

Greek Letters

α	volume shape factor
β	surface shape factor
γ	edge free energy of a growth unit in a step, J
ϵ	growth spiral activity factor
ϱ_c	density of crystal, kg/m³
σ	relative supersaturation
σ_1	BCF model constant
$\bar{\xi}$	retardation factor
Ω	molar volume of a growth unit, m³/mol

LITERATURE CITED

1. Burton, W. K., N. Cabrera and F. C. Frank, Phil. Trans. Roy. Soc. (London), A 243, 299, (1951)

2. Bennema, P., J. Crystal Growth, 1, 278, (1967)

3. Bennema, P., ibidem, 287, (1967)

4. Bennema, P., "The Importance of Surface Diffusion for Crystal Growth from Solution," ibidem, 5, 29, (1969)

5. Cabrera, N., and R.V. Coleman, "Theory of Crystal Growth from the Vapor," in The Art and Science of Growing Crystals, John Wiley & Sons, New York, (1963)

6. Frank, F. C., Growth and Perfection of Crystals, John Wiley & Sons, New York, (1958)

7. Ohara, M., and R. C. Reid, Modeling Crystal Growth Rates from Solution, Prentice-Hall, Englewood Cliffs, (1973)

8. Valcic, A. V., J. Crystal Growth, 30, 129, (1975)

9. Bennema, P., J. Boon, and C. van Leeuwen, "Confrontation of the BCF Theory and Computer Simulation Experiments with Measured (R,σ) Curves," presented at the 4th Congress CHISA'72, Prague, Czechoslovakia, (1972)

10. Alexandru, H.V., "Growth Kinetics of {001} Type Faces of Rochelle-Salt Crystals in Solutions," ibidem

11. Gilmer, G.H., and P. Bennema, J. Crystal Growth, 13/14, 148, (1972)

12. Karpinski, P.H., Mass Crystallization in a Fluidized Bed, Monograph No. 22, TUW Wroclaw, (1981)

13. Budz Jerzy, P. H. Karpinski and Zbigniew

Naruć, AIChE Journal, *30*, 710, (1984)

14. Budz Jerzy, P. H. Karpinski and Zbigniew Naruć, ibidem, *31*, 259 (1985)

15. Karpinski P. H., J. Budz and M. A. Larson, "Influence of Cationic Admixtures on Crystal Growth Kinetics from Aqueous Solution," p. 85 in <u>Proceedings of the 9th Symposium on Industrial Crystallization</u>, Elsevier Science, Amsterdam, (1984)

16. Karpinski P. H. and M.A. Larson, Crystal Res. and Technol., No. 5, 1985

17. Karpinski P. H., Chem. Engng Sci., *35*, 2321, (1980)

18. Karpinski, P. H., Chem. Engng Sci., *40*, 641, (1985)

19. Hosoya, S., M. Kitamura, and T. Miyata, Mineral. J., *9*, No. 3, 147, (1978)

20. Dennis, J. E., "Nonlinear Squares and Equations," in <u>The State of the Art of Numerical Analysis</u>, D. Jacobs, Ed., Academic Press, London, (1977)

21. Nývlt Jaroslav, Otokar Söhnel, Marie Matuchová, and Miroslav Broul, <u>The Kinetics of Industrial Crystallization</u>, Elsevier, Amsterdam, (1985)

22. Bliznakov, G., Kristallografiya, *4*, No. 2, 150, (1959)

THE EFFECT OF Pb(II) AS A TRACE IMPURITY ON THE CRYSTALLIZATION KINETICS OF CaCO$_3$ PRECIPITATION

Robert W. Peters[1] and Tsun-Kuo Chang[2] ■ Environmental Engineering, School of Civil Engineering, Purdue University, West Lafayette, IN 47907

The effect of a trace level contaminant (Pb) was investigated in relationship to the precipitation of CaCO$_3$ in terms of the particle size distribution, precipitation kinetics, particle morphology, and residual hardness and lead concentrations. For the pH range and lead concentrations investigated, the maximum contaminant level (MCL) of 0.05 mg/l can be met for pH in the range of 7.8 to 10.6. The predominant CaCO$_3$ crystalline form observed for MSMPR operation in the presence of lead is that of calcite. The presence of the lead hinders the growth rate of CaCO$_3$ precipitation; it also enhances the nucleation rate, thereby causing a greater decay in the resulting supersaturation. Lower residual calcium concentrations are achieved in the presence of lead than that for the pure CaCO$_3$ sludge. The primary mechanism involved for removal of lead in this study was coprecipitation with CaCO$_3$ resulting in isomorphic inclusion into the CaCO$_3$ crystal structure.

INTRODUCTION

During the past 40 years, disposal of municipal and industrial wastes in landfills has been a widely used practice in the United States. Of 76,000 identified landfill sites, as many as 2000 landfills and dumps present imminent health hazards [1]. When refuse or sludge buried in a landfill comes in contact with water, leachate is produced that can migrate out of the landfill thereby polluting the groundwater. Generally the inorganic constituents in highest concentration in leachate are calcium, magnesium, sodium, potassium, iron, zinc, cadmium, copper, and lead.

Lead is well known for its toxicity; no beneficial effects have been found for human or animal development. Although acute lead poisoning is somewhat rare, chronic lead toxicity is common because lead accumulates in bones and tissues and people are exposed to lead in food, air, and water. Another source of lead is inhaled tobacco smoke.

The general symptoms of lead poisoning are gastrointestinal disturbances, loss of appetite, fatigue, anemia, motor nerve paralysis, and encephalopathy [2]. Because these symptoms are not unique to lead poisoning, it is often misdiagnosed. The major chronic effects are produced in the hematopoietic system, central peripheral nervous systems, and kidneys. A special risk group for lead toxicity is small children in the two to three year old age group, because lead absorption from food and water at that age is 40-50% as compared to 5-10% for adults [3]. Young children experience a greater risk of ingesting lead from other sources, such as lead-containing paints and dust.

[1] Assistant Professor of Environmental Engineering.
[2] Formerly graduate student. Current affiliation: Associate Professor of Agricultural Engineering, National Taiwan University, Taipei, Taiwan. Republic of China.

Lead has been in the United States Public Health Service (USPHS) drinking water standards since 1925 [4], when the limit was first set at 0.1 mg/l. When the USPHS revised its standard in 1962 [5], the lead limit was lowered to 0.05 mg/l, mainly due to the number of sources and the extent of lead exposure had steadily increased over the limit years. EPA used this same limit (0.05 mg/l) in setting the maximum contaminant level for the National Interim Primary Drinking Water Regulations in 1976 [6].

Lead is a relatively minor constituent of the earth's crust, but is widely distributed in low concentrations in sedimentary rock and soils. The average content of lead in soil is reported [7] to be 10 ppm; higher concentrations may occur in some limestone derived soils. The concentration of lead in seawater is reported by Goldberg [8] to be 0.3 μg/l; rainwater and waters in rivers and lakes generally contain much higher lead concentrations.

The principal problem with lead in drinking waters does not result from lead in natural groundwaters or polluted surface waters, but rather from water distribution systems. Kopp and Kroner [9] reported 305 occurrences of lead (19.3%) in 1577 samples from lakes and rivers of the U.S.; the minimum, maximum, and mean values were 0.001, 0.14, and 0.023 mg/l respectively. Durum et al. [10] found that 63% of 730 surface water samples had lead concentrations ranging from 0.001 to 0.05 mg/l. Very few samples had lead concentrations in excess of the maximum contaminant level (MCL) of 0.05 mg/l. Occasionally lead can be found in groundwaters at levels as high as 0.4-0.8 mg/l. Dutt and McCreary [11] reported that 6.5% of 677 water samples collected from Arizona groundwaters had lead concentrations exceeding the MCL, with the highest recorded value being 0.518 mg/l. The 1969 Community Water Supply [12] showed that 37 of 2595 water distribution line samples of 969 public water supply systems contained lead concentrations exceeding the 0.05 mg/l limit. In a more recent survey [13], water samples from 383 households in Boston, Cambridge, and Somerville, Massachusetts had lead concentrations ranging from less than 0.003 mg/l to 1.51 mg/l with the mean of all samples being 0.030 mg/l.

BACKGROUND

The removal of sparingly soluble carbonate hardness can be accomplished through the use of precipitation techniques of the lime-soda ash water softening process. Precipitation of heavy metals through the addition of lime is also a well established technology (14,15). The by-products of water softening by precipitation methods are calcium carbonate and magnesium hydroxide sludge. These materials provide a good media for adsorption of both organic and inorganic substances in water (16,17). Thus, lime softening for removal of hardness should have the ability for removal of heavy metals by both coprecipitation and adsorption.

The principal source of information to date on lime softening for treatment of trace level inorganic removal is laboratory and pilot plant studies conducted by the Drinking Water Research Division (DWRD) of the U.S. EPA at the Environmental Research Center in Cincinnati, Ohio (16). Jar-test studies with spiked well water showed lime and excess lime softening to be more effective than iron or alum coagulation for lead removal. Greater than 99% removals for an initial lead concentration of 0.15 mg/l were achieved throughout the pH range of 8.8 to 11.0. Reductions exceeding 99% were also obtained at pH 9.4-9.6 for initial lead concentrations of 10 mg/l. One pilot plant test run conducted at pH 9.5 showed the lead removal to be in good agreement with the jar test data. The large amount of $CaCO_3$ and $Mg(OH)_2$ floc developed was speculated as accounting for the high lead removal efficiencies.

To date, most of the lime precipitation studies focus on hardness removal or heavy metal removal alone. In most applications however, pure solutions are rare. The effects of impurities on both the crystallization kinetics behavior of $CaCO_3$ and the coprecipitation/adsorption behavior of heavy metals by softening processes are very important from both hardness removal and heavy metals removal points of view. Since lead is a common industrial pollutant, this research is aimed at using the lime-soda ash water softening process in the presence of trace amounts of lead to evaluate the effect of lead as an impurity on the crystallization kinetics, particle size distributions, particle morphology, and removal efficiency of hardness as well as to understand the mechanisms involved in heavy metal removal in this process.

Prior to this study, several studies have been conducted on the effects of selected heavy metals on calcium carbonate precipitation. Peters and Stevens (18) studied the effect of iron added as an impurity to the calcium carbonate system. Iron can be removed through the use of the lime-soda ash water softening process forming insoluble iron precipitates which are amorphous in nature. The removal of iron was strongly pH dependent. The iron markedly affects the amount of wall deposition in the continuous precipitation system. The presence of iron caused a slight inhibition of the growth rate, but greatly enhanced the nucleation rate. The crystal growth rate of the calcium carbonate-magnesium hydroxide precipitate decreased with increasing iron concentration. The iron was speculated as providing potential sites for nucleation. The growth inhibition contributed to the formation of a "bundle" or "dumbbell-shaped" crystal structure for the aragonite precipitates.

In a previous study by the authors (19), cadmium was found to be effectively removed from both synthetic hard water systems and groundwaters spiked with cadmium. For the pH range investigated, the MCL was able to be met for pH in the range of 7.3 to 11.0. Low cadmium concentrations ($Cd_i \leq 1.0$ mg/l) had no effect on the batch precipitation of $CaCO_3$, at least in terms of the residual calcium concentrations obtained. High initial cadmium concentrations (~5 mg/l) interfered with the precipitation of $CaCO_3$, suggesting a threshold concentration of ~2 mg/l for a noticeable inhibition on $CaCO_3$ crystal growth. With 5.0 mg/l of cadmium initially present, the residual calcium level increased by 20-30%. Calcite was the only crystal morphological form observed for the continuous $CaCO_3$ precipitation. In the absence of impurities, the kinetically favored metastable aragonite crystalline form predominates. The presence of cadmium severely inhibited aragonite growth, causing the thermodynamically favored calcite form to prevail. No effect was experienced on the removal of magnesium due to the presence of cadmium. For the continuous precipitation studies, the residual calcium concentration increased ~30-40 mg/l in the presence of cadmium again indicating the inhibitory effect of cadmium of $CaCO_3$ precipitation. Chang and Peters (19) felt such behavior explains why almost all municipalities employing lime softening operations have calcite as the predominant crystalline $CaCO_3$ form, due to the fact that most groundwaters contain trace levels of inorganic contaminants. Removal of cadmium was greatest in the early stages of the runs due to the larger adsorptive capacity. Removal of cadmium is due primarily to physical adsorption onto the $CaCO_3$ sludges.

Research has also been conducted on the calcite-metal interaction in soils. Leeper (20) postulated that, in calcarous soils, calcite may be an important adsorbent of heavy metals. Faust and Schultz (21) observed that calcite adsorbs Cd, Cu, Pb, and Zn from water in both batch and columnar applications.

OBJECTIVES

The objectives of our research program are summarized below:

1. Set an operational criteria for a water treatment plant having both hardness and heavy metal problems.

2. Measure the efficiency of heavy metal removal by the lime-soda ash water softening process.

3. Measure the effects of hardness removal in the presence of heavy metals.

4. Measure the precipitation kinetics through a population balance analysis for the calcium carbonate system in the presence of heavy metals.

5. Monitor the crystal morphology to determine the adsorption and coprecipitation phenomena.

6. Use real groundwaters and surface waters (spiked with heavy metals) to evaluate the kinetic model.

The goal thus is to identify the significance of each mechanism as it relates to heavy metal removal from drinking waters.

Studies conducted in our research program are summarized below:

1. Determination of the heavy metal removal efficiency for single

and multiple metal systems (with no hardness present) in both batch and continuous crystallizers, as functions of pH, supersaturation, and their concentration.

2. Conducting experiments on the $CaCO_3-Mg(OH)_2$ system (with no heavy metals present) to determine the precipitation kinetics and hardness removal for the pure hardness system.

3. Conducting experiments on pH adjusted systems to which lime sludge has been added to study the adsorption of heavy metals onto aragonite and calcite. These studies are performed for various pH levels, suspended solids concentrations, and heavy metal concentrations to determine the isotherms for each metal.

4. Conducting experiments using a combination of several heavy metals to determine the interactions of these species.

5. Using real groundwater and surface waters (spiked with heavy metals) to evaluate the kinetic model.

Our research project is thus divided into several phases:

1. Precipitation only of heavy metals by hydroxide and carbonate precipitation.

2. Adsorption of heavy metals onto lime softening sludges.

3. Coprecipitation and adsorption of heavy metals in the lime-soda ash water softening process.

This particular paper focuses its attention on the removal of lead by coprecipitation/adsorption in lime softening operations.

EXPERIMENTAL PROCEDURE

Continuous Coprecipitation/Adsorption Study

The flow diagram for the experimental system is shown in Figure 1. Feed solutions of hardened water (containing heavy metals), lime, and sodium hydroxide are pumped through 0.45 μm filters, passed through constant temperature baths maintained at 25.0 $\pm$ 0.1°C, into the reactor. The lime and sodium hydroxide solutions are prepared with reagent grade $Ca(OH)_2$ and NaOH, respectively. The artificially hardened water is prepared by adding reagent grade calcium chloride to deionized water followed by addition of a slight excess of sodium bicarbonate to convert the calcium hardness to carbonate hardness. A concentrated lead solution (1000 mg/l) is added to this feed solution to obtain the desired initial lead concentration in the reactor. The feed stream volumetric flow rates are adjusted to maintain 4.0 l of magma in the crystallizer at the desired residence (detention) time (of 15.0, 20.0, or 30.0 minutes). Table 1 summarizes the nominal experimental conditions employed for these continuous precipitation experiments. Steady state operation is generally achieved after 10-11 residence times. Most of the data are collected between 12 and 20 residence times. The data collected include: hardness titrations, heavy metal concentrations, temperature and flow rate measurements, conductivity, alkalinity and pH measurements, suspended solids concentrations, and particle size distributions. The particle size distributions are determined using a Coulter Counter Model TA-II. The metal analyses were analyzed using an Allied Video 12 AA/AE spectrophotometer (manufactured by Instrumentation Laboratory, Inc.).

RESULTS AND DISCUSSION

Preliminary Batch Precipitation Experiments

To determine the effect of lead on the precipitation of calcium carbonate, batch $CaCO_3$ precipitation experiments were first performed in the presence of lead for various pH conditions and initial calcium concentrations. The results of this batch coprecipitation/adsorption study are summarized in Table 2 and in Figures 2 and 3. Figure 2 plots the residual calcium concentration for various initial calcium and lead concentrations as a function of pH for these batch precipitation conditions. The solid symbols in the figure represent the residual calcium concentration obtained for the $CaCO_3$ without any lead present. As shown in Table 2 and Figure 2, extremely low residual calcium concentrations (~3.0 mg/l) can be achieved independent of the initial calcium concentration used; all three initial concentrations fall on the same calcium concentration -pH curve. For pH > 10.3, the residual calcium concentration is generally less than 4 mg/l. Note that these conditions apply to the pure $CaCO_3$ system in the absence of impurities.

Using initial calcium concentrations of 150, 250, and 350 mg/l as $CaCO_3$ and an initial lead concentration of 1.0 and 5.0 mg/l, the coprecipitation/adsorption phenomena for removal of lead was studied. The results of these experiments are likewise summarized in Table 2 and Figures 2 and 3. At low initial lead concentrations (~1.0 mg/l), the residual calcium concentration is virtually unaffected by the presence of lead as shown in Figure 2. Figure 3 shows both the residual calcium and lead concentrations obtained as a function of pH. Using all three initial calcium concentrations, the MCL can be met for pH in the range of 7.7 to 11.0. This result is similar to that obtained by Sorg et al. [16] who claimed the MCL could be met for pH in the range of 8.8 to 11.0. The results obtained from our study show that the MCL can be met 1.1 pH units lower than that obtained by Sorg et al., resulting in a lower treatment cost. Figure 3 also shows that the pH range capable of meeting the MCL at the higher lead concentration is reduced slightly, namely pH in the range of 7.8 to 10.6.

Referring back to Figure 2, considerably higher residual calcium concentrations are obtained in the presence of 5.0 mg/l lead. The higher initial lead concentration hindered the calcium carbonate precipitation. This is likely due to the competition for carbonate by both Ca and Pb along with surface adsorption of lead (with incorporation into the calcite crystal structure) on the calcite surface thereby reducing the number of growth sites.

Continuous Precipitation Study

A series of continuous precipitation experiments conducted in the presence of lead were likewise performed. The nominal operating conditions for these experiments are summarized in Table 1. The residual hardness and lead concentrations from these experiments are summarized in Tables 3 and 4, respectively, while Table 5 summarizes the alkalinity measurements for each experimental run. The mean effluent concentrations are summarized for each series of runs in Table 6.

The residual hardness level is dramatically affected by the presence of impurities such as lead, as shown in Table 3. In the absence of impurities, at pH 10.3, the residual calcium concentration is approximately 40-55 mg/l. In the presence of lead, lower residual calcium concentrations are obtained ranging from 25-48 mg/l. Lead fits nicely into the crystal lattice of calcite and becomes incorporated into the crystal lattice. Due to this fact along with a higher nucleation rate, the supersaturation is quickly relieved. Note that the initial lead concentrations were designed to be 5.0 mg/l; however as shown in Table 4, the initial lead concentration after passing through the filter prior to entry into the reactor was such that the effective concentration entering the reactor was 1.22 mg/l, probably due to formation of $Pb(OH)_2$ flocs. Since the initial lead concentration was present in very low concentrations, no distinguishable peak could be used to identify the lead compounds (which were masked by the calcite peaks) using x-ray diffraction analysis techniques.

Series VII and X listed in Table 3 employed higher initial hardness levels (250 mg/l Ca^{++} and 65 mg/l Mg^{++} as $CaCO_3$). The calcium level is affected by the presence of Mg^{++}. Since magnesium is an inhibitor of $CaCO_3$ precipitation, higher residual calcium concentrations are obtained. The presence of lead increased the removal efficiency for magnesium hardness. Comparing series VII and X, the residual magnesium hardness levels are 19-41 versus 40-54 mg/l as $CaCO_3$, respectively. Table 5 shows the lead had a negligible influence on the alkalinity of the system.

Table 4 shows the lead was very effectively removed with the $CaCO_3$–$Mg(OH)_2$ precipitations due to a coprecipitation/adsorption phenomena. Lower residual calcium concentrations were obtained when lead was present as an impurity. The residual lead concentration for all samples from series III and VII were below the usual detection limit (0.01 mg/l). For this research, the Smith-Hieftje background correction method was employed to correct for some spectral overlap interferences. With this method, the detection limit for lead is 0.003 mg/l. Several explanations are possible to account for these high removal efficiencies of lead documented in Table 4. Possible explanations include:

1. Lead is isomorphically included into the calcium carbonate crystals during $CaCO_3$ precipitation.
2. Very low initial lead concentrations were used (1.22 mg/l).
3. A more sensitive detection limit was employed.

Particle Size Distribution/Population Balance Theory

Industrial water softening operations closely approximate the operation of a mixed suspension mixed product removal (MSMPR) crystallizer. The MSMPR analysis developed by Randolph and Larson (22) is used in this study. Use of the population balance for the MSMPR crystallizer along with the assumption of size independent growth (22) yields:

$$n = n° \exp(-L/G\tau) \quad (1)$$

Due to the low suspended solids concentrations involved in this research, it is more accurate to plot the total number of crystals in the size range L to ∞:

$$N(L,\infty) = \int_L^\infty n \, dL = \int_L^\infty n° \exp(-L/G\tau)dL$$
$$= n°G\tau \exp(-L/G\tau) \quad (2)$$

The nucleation rate and dominant particle size are defined by the equations below:

$$B° = n° G \quad (3)$$
$$L_D = 3G\tau \quad (4)$$

Figure 4 shows a plot of the particle size distribution (PSD) achieved for precipitation of $CaCO_3$ in the presence of 1.22 mg/l Pb at pH 10.3 for a reactor detention time of 15.0 minutes from Run No. Pb-1 of Series III. For the particular case shown in the figure, the slope of the line is $(-1/G\tau) = 0.1768$. Knowing this information, the following parameter values are obtained:

G = 0.377 μm/min.
$n°$ = 17000 no./ml–μm
$B°$ = 6400 no./ml-min.
L_D = 17.0 μm

The correlation coefficient for this linear regression is -0.998 indicating the MSMPR model provides a very adequate representation of this process. The results of the various particle size distributions achieved in this MSMPR study are summarized in Table 7. This table shows that the presence of lead hinders the growth rate of $CaCO_3$ precipitation while enhancing the nucleation rate. The net result is the formation of smaller particles with the presence of lead than that obtained in the absence of lead, making solid-liquid separations more difficult. Also noteworthy from Table 7 is that the predominant crystalline form in the absence of lead was the aragonite form whereas calcite was the predominant crystalline form in the presence of lead. Such behavior explains why almost all municipalities employing lime softening operations have calcite as the predominant crystalline $CaCO_3$ form, due to the fact that most raw waters contain trace levels of inorganic contaminants.

The relationship of the supersaturation driving forces to nucleation and growth rates is of considerable importance. At constant temperature, the nucleation and growth rates can be modeled with simple power law models, which, when combined, yield:

$$B° = k_N G^i \quad (5)$$
$$n° = k_N G^{i-1} \quad (6)$$

where: k_N is the kinetic rate constant relating the nucleation rate to the growth rate;

i is the kinetic order.

The values of k_N and i may be determined experimentally from a series of runs performed at different supersaturation levels, accomplished in this case by varying the reactor detention time. Table 8 summarizes the precipitation kinetics for the continuous precipitation of $CaCO_3$ both in the absence and in the presence of trace levels of lead. The nucleation rate-growth rate relationships for the various series of runs also shown in Figure 5. Tables 7 and 8 indicate that the presence of both lead and magnesium hinders the growth rate of calcium carbonate while enhancing the nucleation rate. The presence of lead decreases the value of i and increases k_N resulting in a less favorable size distribution (smaller particles) than the same system in the absence of lead. The presence of magnesium along with a slight increase in pH resulted in both i and k_N increasing (both in the presence and in the absence of lead).

Particle Morphology

The three major crystalline forms of calcium carbonate are calcite, aragonite, and vaterite (23,24). Calcite crystals are of rhombic shape, aragonite needle-like, and vaterite disk-like or spherulitic (24). Calcite is the most stable thermodynamically and aragonite is a metastable kinetically favored crystalline form. Calcite and aragonite are the most common crystalline forms found in nature (24).

A sample of crystals is obtained from a typical experiment, filtered, and dried, and studied under a scanning electron microscope. The crystal morphology from various experimental runs was studied to observe the effect on the crystal habit by changing the reactor detention time, the pH level, and the presence of various impurities (Cd, Pb, or Zn) on the $CaCO_3$ system. Table 7 lists the predominant calcium carbonate crystalline forms found in the MSMPR study.

Figure 6, with a magnification of 4000x, shows a close-up of the crystals obtained in the batch $CaCO_3$ precipitation conducted in the absence of any impurities. The crystals are from the suspension collected at the end of 4.0 hours of reaction time for an initial hardness level of 250 mg/l as $CaCO_3$ and a pH of 10.3. Only the calcite morphological form is observed under these conditions.

In contrast to the previous figure, Figure 7 shows a close-up of a calcium carbonate crystal from a previous study (25). The aragonite particle is magnified 2000x. The sample was collected from the suspension collected at 17.4τ for a run conducted at pH 10.3 with a reactor detention time of 19.85 minutes. The initial hardness level was 246.1 and 65.0 mg/l as $CaCO_3$ for calcium and magnesium, respectively. The dendritic structure of the aragonite crystal is typical of large driving forces (concentration gradients).

Figure 8 shows the crystal morphology characteristics of the particles collected from Series III for the conditions: pH = 10.3, initial calcium concentration = 250 mg/l as $CaCO_3$, and an initial lead concentration of 1.22 mg/l. For the case shown in Figure 8, the reactor detention time is 15.0 minutes. The presence of lead promotes the nucleation rate, causing the supersaturation to be quickly relieved; low residual calcium concentrations are obtained. This low supersaturation condition favors the formation of calcite; in all four experimental runs of Series III calcite was the only crystal morphological form found. Figure 9 shows the same type of behavior; for this case, the reactor detention time was 30.0 minutes. Both figures show a well defined calcite structure.

For Series VII employing conditions of pH = 10.8, initial calcium concentration = 250.0 mg/l as $CaCO_3$, initial magnesium concentration = 65.0 mg/l, and an initial lead concentration = 1.22 mg/l, nicely formed calcite particles are still the predominant $CaCO_3$ crystalline form, although aragonite can be found in lesser quantities. For the particular case shown in Figure 10, the reactor detention time is 15.0 minutes. The particle size analyses not only indicated the number of particles increased with addition of magnesium, but also that the particle size decreased. The hardness removal data presented in Table 3 also shows the removal efficiency of magnesium improved with the addition of lead.

Comparison with $Pb(OH)_2$ Precipitation

Many of the metals precipitate as insoluble hydroxides if the pH is increased by adding lime or caustic soda. However, certain amphoteric elements redissolve at high pH. The solubility of the metal hydroxide is defined by the solubility product (K_{SP}) and equations relating the solid metal ion or metal hydroxide complexes to the insoluble precipitates. The equations describing the solubility of lead hydroxide are listed below:

Solubility Product

$$Pb(OH)_2(s) \rightleftharpoons Pb^{++} + 2\,OH^-$$

$$K_{so} = [Pb^{++}][OH^-]^2 = K_{SP} \qquad (7)$$

Complex Formation

$$Pb^{++} + OH^- \rightleftharpoons PbOH^+$$

$$K_{s1} = \frac{[PbOH^+]}{[Pb^{++}][OH^-]} \qquad (8)$$

$$PbOH^+ + OH^- \rightleftharpoons Pb(OH)_2$$

$$K_{s2} = \frac{[Pb(OH)_2^o]}{[PbOH^+][OH^-]} \qquad (9)$$

$$Pb(OH)_2^o + OH^- \rightleftharpoons Pb(OH)_3^-$$

$$K_{s3} = \frac{[Pb(OH)_3^-]}{[Pb(OH)_2^o][OH^-]} \qquad (10)$$

Water Ionization

$$H_2O \rightleftharpoons H^+ + OH^-$$

$$K_w = [H^+][OH^-] \qquad (11)$$

Taking the logarithmic form of the above equations and substituting $\log[OH^-] = pH - 14$ gives the following equations:

$$\log[Pb^{++}] = \log K_{so} - 2pH + 28 \qquad (12)$$

$$\log[PbOH^+] = \log K_{s1} + \log K_{s0} - pH + 14 \quad (13)$$

$$\log[Pb(OH)_2^o] = \log K_{s2} + \log K_{s1} + \log K_{s0} \quad (14)$$

$$\log[Pb(OH)_3^-] = \log K_{s3} + \log K_{s2} + \log K_{s1} + \log K_{s0} + pH - 14 \quad (15)$$

$$\log[OH^-] = pH - 14 \quad (16)$$

where the K_{si}'s are the differential ionization constants. In addition to the above equations a mass balance for the soluble lead must be maintained:

$$C_T = [Pb^{++}] + [PbOH^+] + [Pb(OH)_2^o] + [Pb(OH)_3^-]$$

Using the data given by Patterson et al. (26), the solubility diagrams are constructed as shown in Figure 11. Note that the fresh precipitates tend to be more soluble than aged precipitates. This behavior is generally indicative of a phase transformation from an amorphous kinetically favored precipitate to a more ordered crystalline form. By comparison, the results presented in Table 2 from the batch precipitation study are presented on the same figure. Note that the residual lead concentration is 3-4 orders of magnitude less than that predicted from $Pb(OH)_2$ solubility considerations due to coprecipitation/adsorption phenomena. The x-ray diffraction pattern data, the scanning electron microscope photographs, and the residual heavy metals concentration all suggest that the major mechanism responsible for removal of lead from solution in the lime-soda ash water softening process is isomorphic inclusion.

SUMMARY

The removal of lead from hard water supplies with the lime-soda ash water softening process has been studied. Lead can be very effectively removed contaminated groundwaters using conventional lime softening operations. For the pH range and lead concentrations investigated, the MCL can be met for pH in the range of 7.8 to 10.6. Trace level lead concentrations ($Pb_i \leq 1.0$ mg/l) have no effect on the residual calcium concentration obtained from batch $CaCO_3$ precipitation, while higher initial lead concentrations ($Pb_i \sim 5.0$ mg/l) interfered with the precipitation of $CaCO_3$; the residual calcium hardness increased.

The predominant crystalline form of $CaCO_3$ precipitated in the presence of lead is of the calcite form. For the same conditions in the absence of impurities, aragonite is the major calcium carbonate precipitate polymorph. Since most water supplies contain trace levels of impurities, most municipal lime softening operations result in calcite precipitation. In general, lead can be very effectively removed by the lime-soda ash water softening process without sacrificing the hardness removal efficiency. The removal of magnesium hardness is enhanced by the presence of lead for MSMPR operation at pH 10.8. The presence of lead also decreased the residual calcium concentration resulting in a lower supersaturation. The presence of lead caused a decrease in the growth rate of the $CaCO_3$ precipitates of ~45% and an increase in the nucleation rate of ~4x, causing much smaller particles to be formed. The dominant particle size decreased from ~33 μm in the absence of lead to ~16 μm with lead present. The xray diffraction pattern data, the scanning electron microscope photographs, and the residual heavy metal concentration data all suggest the primary mechanism involved for removal of lead in lime softening operations is isomorphic inclusion.

ACKNOWLEDGEMENTS

The authors wish to acknowledge the support of the School of Civil Engineering at Purdue University enabling this research to be performed. A special word of thanks goes to Mrs. Janet E. Lovell in performing the scanning electron microscope and xray diffraction analyses.

NOMENCLATURE

B^o	Particle nucleation rate, no./ml-min.
Ca^{++}	Calcium concentration, mg/l as $CaCO_3$
$CO_3^=$	Carbonate alkalinity, mg/l as $CaCO_3$
C_T	Total soluble lead concentration, moles/l
G	Particle growth rate, μm/min.
H^+	Hydrogen ion concentration, moles/l
HCO_3^-	Bicarbonate alkalinity, mg/l as $CaCO_3$
i	Kinetic exponent relating nucleation rate to growth rate
k_N	Kinetic rate constant relating nucleation rate to growth rate
K_{si}	Solubility constant of the ith lead hydroxide complex, moles/l
K_{sp}	Solubility product, mole^3/l^3
L	Particle size, μm
L_D	Dominant particle size, μm
Mg^{++}	Magnesium concentration, mg/l as $CaCO_3$
MCL	Maximum contaminant level
MSMPR	Mixed suspension mixed product removal
n	Population density at size L, no./ml-μm
n^o	Nuclei population density, no./ml-μm
N	Cumulative number of particles per ml
ND	Not detected
OH^-	Hydroxide alkalinity, mg/l as $CaCO_3$
pH	-log $[H^+]$
Pb^{++}	Lead concentration, mg/l
PSD	Particle size distribution
r	Correlation coefficient
τ	Crystallizer detention time, min.

LITERATURE CITED

1. "Current Developments", *Environ. Reporter*, 269, (1980).

2. Goodman, L. S., and A. Gilman, *The Pharmacological Basis of Therapeutics*, MacMillan Co., London and Toronto, (1970).

3. Drinking Water and Health, Recommendations of the National Academy of Science, *Fed. Reg.*, 42(132): 35764, (July 11, 1977).

4. Report of Advisory Committee on Official Water Standards, *Pub. Health Reports*, 40(15): 693, (April 10, 1975).

5. Drinking Water Standards, U.S. Public Health Service, Department of Health, Education and Welfare, PHY Publ. No. 956, (1962).

6. EPA, "National Interim Primary Drinking Water Regulations", *Water Prog. Fed. Reg.*, *40*: 148, (Dec. 24, 1975).

7. Bowen, H. J. N., *Trace Elements in Biochemistry*, Academic Press, New York, NY, (1966).

8. Goldberg, D. E., "Chemistry - The Oceans as a Chemical System", in *Composition of Sea Water; Comparative, and Descriptive Oceanography*, Vol. 2, M. N. Hill, ed., Wiley Interscience, New York, NY, (1963).

9. Kopp, J. F., and R. C. Kroner, *Trace Metals in Waters of the United States*, U.S. Dept. of the Interior, Fed. Water Pollut. Control Admin., Washington, D.C., (1970).

10. Durum, W. H., J. D. Hem, and S. C. Heidel, "Reconnaissance of Selected Minor Elements in Surface Waters of the United States", U.S. Geol. Survey Circ. No. 643, 49, (1971).

11. Dutt, G. R., and T. W. McCreary, "The Quality of Arizona's Domestic, Agricultural and Industrial Wastes", Report No. 256, Agricultural Experiment Station, University of Arizona, Tucson, AZ, (Feb. 1970).

12. McCabe, L. J., J. M. Symons, R. D. Lee, and G. G. Robeck, "Survey of Community Water Supply Systems", *J. Am. Water Works Assoc.*, *62*(11): 670-687, (1970).

13. Karalekas, P. C., *et al.*, "Lead and Other Trace Metals in Drinking Water in the Boston Metropolitan Area", *Proc. Water Works Assoc., 95th Annual Conf.*, Minneapolis, MN, (1975).

14. Brantner, K. A., and E. J. Cichon, "Heavy Metals Removal: Comparison of Alternative Precipitation Processes", *Proc. 13th Mid-Atlantic Industrial Waste Conf.*, *13*: 43-50, (1981).

15. Maruyama, T., S. A. Hannah, and J. M. Cohen, "Metal Removal by 'Physical and Chemical Treatment Processes", *J. Water Pollut. Control Fed.*, *47*(5): 962-975, (1975).

16. EPA, *Manual of Treatment Techniques for Meeting the Interim Primary Drinking Water Regulations*, EPA 600/8-77-005, (1977).

17. Randtke, S. J., C. E. Thiel, M. Y. Liao, and C. N. Yamaya, "Removing Soluble Organic Contaminants by Lime - Softening", *J. Am Water Works Assoc.*, *74*(4): 192-202, (1982).

18. Peters, R. W., and J. D. Stevens, "Effect of Iron as a Trace Impurity on the Water Softening Process", *AIChE Symp. Series, Nucleation, Growth, and Impurity Effects in Crystallization Process Engineering*, *78*(215): 46-67, (1982).

19. Chang, T. -K., and R. W. Peters, "Removal of Cadmium from Contaminated Groundwaters through Coprecipitation and Adsorption in Lime Softening Operations", *Proc. 17th Mid-Atlantic Industrial Waste Conf.*, *17*: 455-474, (1985).

20. Leeper, G. W., *Plant Physiol.*, *1*: 13, (1952).

21. Faust, S. D., and C. M. Schultz, "The Efficacy of Removal of Heavy metals from Water by Calcite", *J. Environ. Sci. Health*, *A18*(1): 95-102, (1983).

22. Randolph, A. D., and M. A. Larson, *Theory of Particulate Processes*, Academic Press, New York, NY, (1971).

23. Dedek, J., *Le Carbonate aux Chaux*, Louvain: Librairie Universitaire, (1966).

24. Wray, J. L., and F. Daniels, "Precipitation of Calcite and Aragonite", *J. Am. Chem. Soc.*, *79*: 2031-2034, (1957).

25. Peters, R. W., L. D. Swinney, and J. D. Stevens, "Precipitation Kinetics of Calcium Carbonate and Magnesium Hydroxide in a Scaling System", *AIChE Symp. Series, Water - 1980*, *77*(209): 49-66, (1981).

26. Patterson, J. W., H. E. Allen, and J. J. Scala, "Carbonate Precipitation for Heavy Metals Pollutants", *J. Water Pollut. Control Fed.*, *49*(12): 2397-2410, (1977).

27. Peters, R. W., P. -H. Chen, and T. -K. Chang, "$CaCO_3$ Precipitation Under MSMPR Conditions", *Industrial Crystallization 84*, 309-316, *Proc. 9th Symposium on Industrial Crystallization*, The Hague, The Netherlands, September 25-28, (1984).

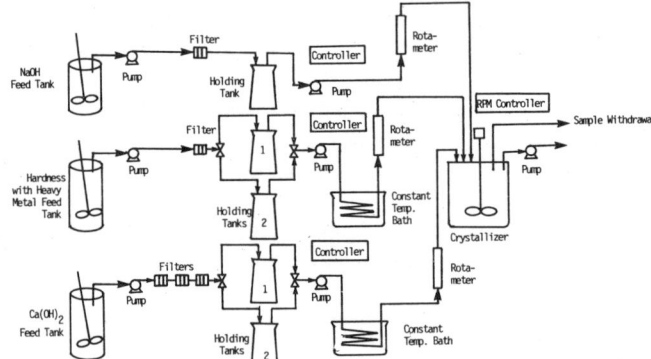

Figure 1. Flow diagram of experimental equipment.

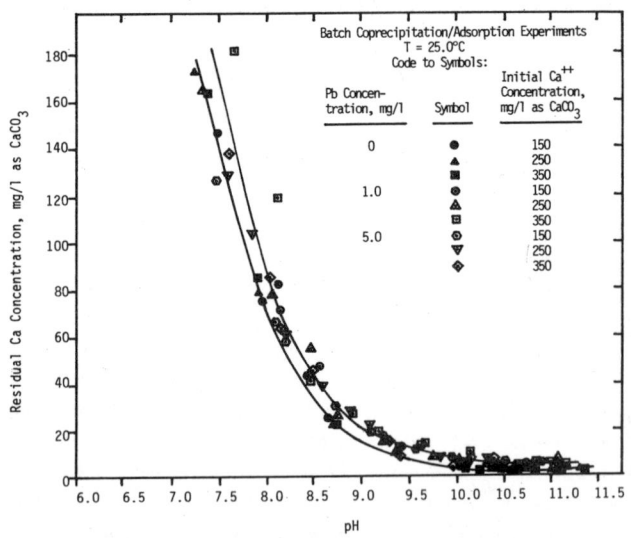

Figure 2. Effect of pH and Pb concentration on the residual calcium concentration obtained from batch precipitation of $CaCO_3$.

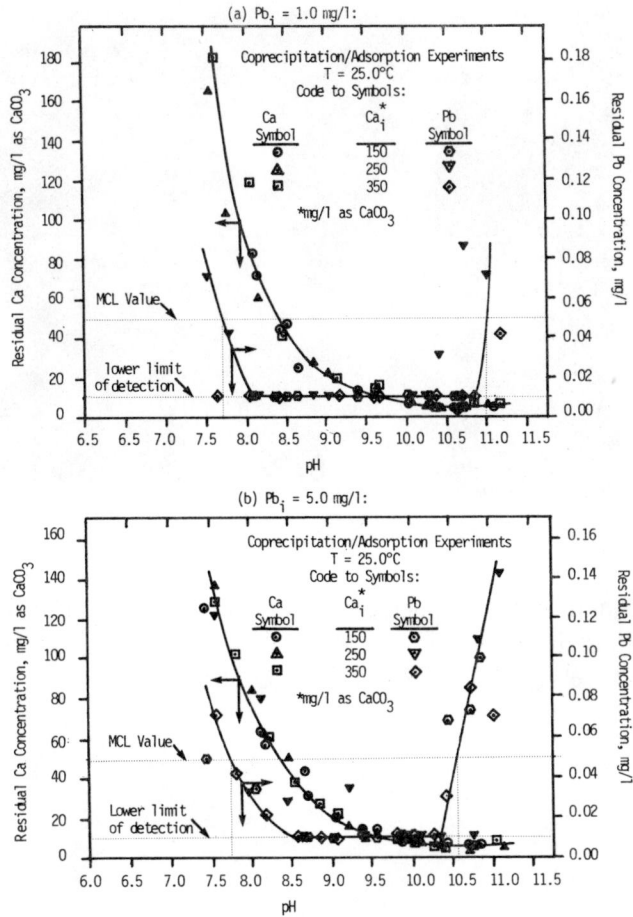

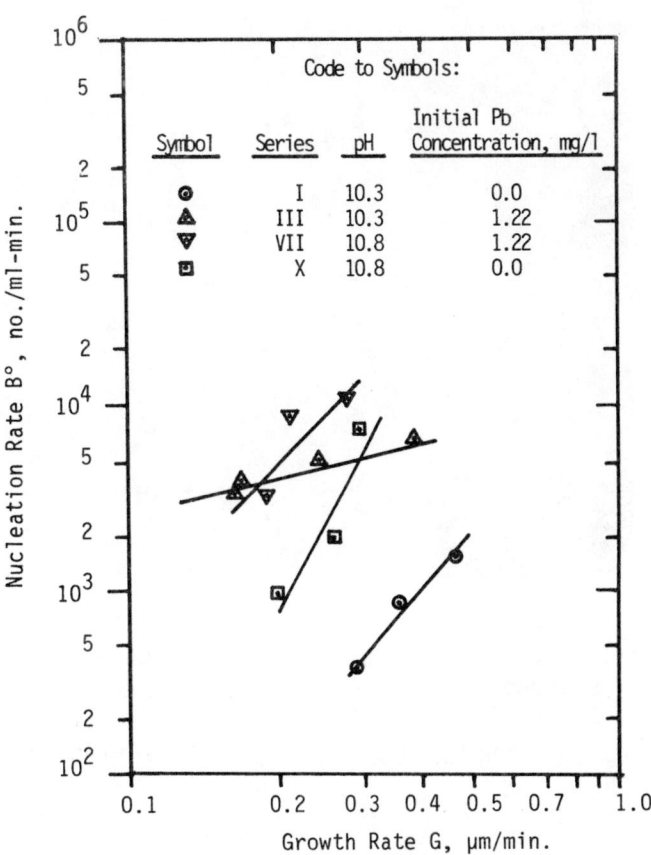

Figure 3. Residual calcium and lead concentration obtained from batch coprecipitation/adsorption experiments using: (a). Pb_i = 1.0 mg/l (b). Pb_i = 5.0 mg/l.

Figure 5. Nucleation rate—growth rate relationships for continuous $CaCO_3$ precipitation conducted in the presence of lead.

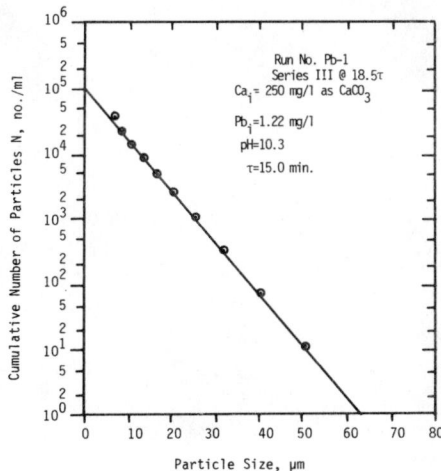

Figure 4. Particle size distribution for continuous $CaCO_3$ precipitation conducted in the presence of 1.22 mg Pb/l at pH 10.3 for Run No. Pb-1 from Series III employing a 15.0 minute detention time.

Figure 6. Photomicrograph of $CaCO_3$ aggregate magnified 4000x obtained from batch precipitation of $CaCO_3$ in the absence of any impurities at pH 10.3.

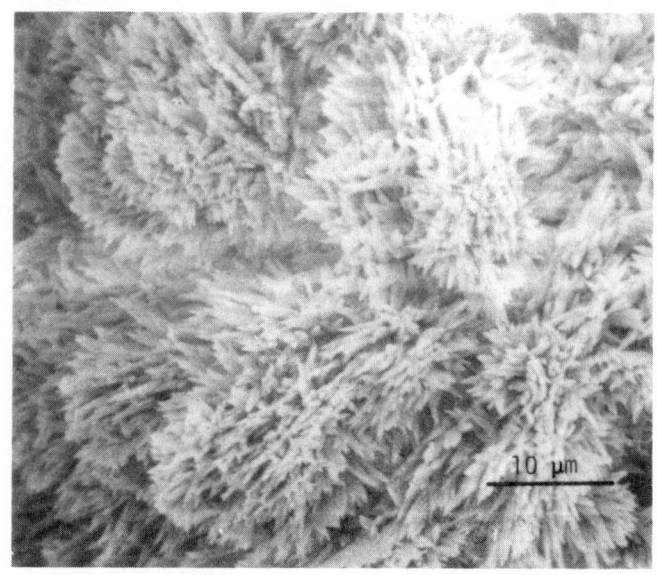

Figure 7. Photomicrograph of surface of aragonite crystals magnified 2000x from suspension collected at 17.4 τ from Run 5 (pH = 10.3, Ca_i = 246.1 mg/l as $CaCO_3$, Mg_i = 65.0 mg/l as $CaCO_3$, Fe_i = 0.0 mg/l, τ = 19.85 min.). (25).

Figure 9. Photomicrograph of particles magnified 3000x from Run No. Pb-3 of Series III (pH = 10.3, Ca_i = 250 mg/l as $CaCO_3$, Pb_i = 1.22 mg/l, τ = 30.0 min.).

Figure 8. Photomicrograph of particles magnified 3000x from Run No. Pb-2 of Series III (pH = 10.3, Ca_i = 250 mg/l as $CaCO_3$, Pb_i = 1.22 mg/l, τ = 20.0 min.).

Figure 10. Photomicrograph of particles magnified 3000x from Run No. Pb-5 of Series VII (pH = 10.8, Ca_i = 250 mg/l as $CaCO_3$, Mg_i = 65 mg/l as $CaCO_3$, Pb_i = 1.22 mg/l, τ = 15.0 min.).

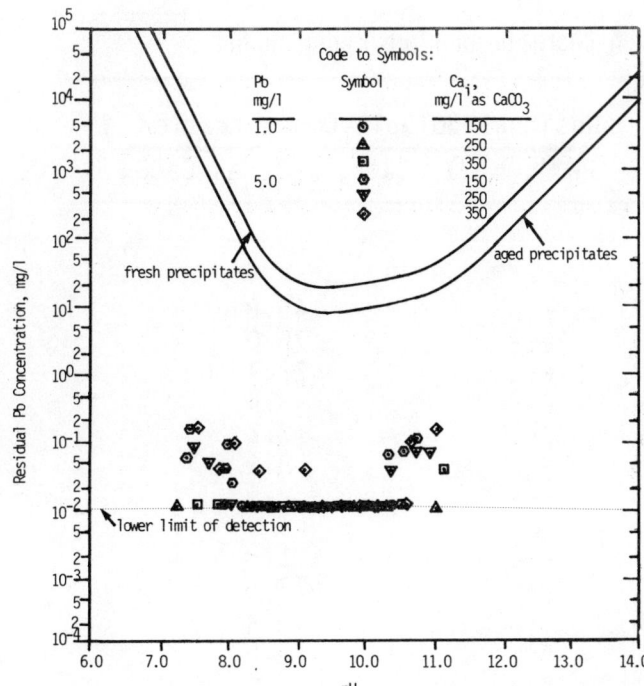

Figure 11. Comparison of residual lead concentrations obtained from batch coprecipitation/adsorption experiments with the theoretical $Pb(OH)_2$ solubility as a function of pH.

Table 1. Nominal MSMPR experimental conditions.

Series	Run No.	Hardness, mg/l as $CaCO_3$			pH	Pb^{++} Concentration, mg/l	Detention Time τ, min.
		Ca^{++}	Mg^{++}	Total			
I	P-1	250	0	250	10.3	0	20.0
	P-2	250	0	250	10.3	0	30.0
	P-3	250	0	250	10.3	0	40.0
III	Pb-1	250	0	250	10.3	5.0	15.0
	Pb-2	250	0	250	10.3	5.0	20.0
	Pb-3	250	0	250	10.3	5.0	30.0
	Pb-4	250	0	250	10.3	5.0	20.0
V	Com-1	250	0	250	10.3	1.0**	20.0
VII	Pb-5	250	65	315	10.8	5.0	15.0
	Pb-6	250	65	315	10.8	5.0	20.0
	Pb-7	250	65	315	10.8	5.0	30.0
X	M-1*	250	65	315	10.8	0	20.0
	M-2*	250	65	315	10.8	0	30.0
	M-3*	250	65	315	10.8	0	40.0

* Study performed by Pu-Hua Chen (27).
**1.0 mg/l each of Cd, Zn, and Pb.

Table 2. Residual concentrations from batch coprecipitation/adsorption studies.

Initial Ca^{++} Concentration, mg/l as $CaCO_3$	Final pH	Residual Soluble Concentration	
		Pb^{++}, mg/l	Ca^{++}, mg/l as $CaCO_3$

$Pb_i = 0$ mg/l:

150	7.47	---	145.8
	7.92	---	75.0
	8.72	---	22.7
	10.25	---	3.4
	10.56	---	3.2
	11.08	---	3.1
	11.33	---	2.9
250	7.27	---	171.8
	7.91	---	77.8
	8.68	---	23.4
	10.10	---	4.6
	10.53	---	3.6
	10.77	---	3.3
	11.00	---	3.1
	11.30	---	3.0
350	7.38	---	163.3
	7.88	---	84.7
	8.70	---	23.3
	10.26	---	3.3
	10.63	---	3.6
	10.83	---	3.1
	11.10	---	3.1
	11.32	---	3.0

$Pb_i = 1.0$ mg/l:

150	8.10	ND	81.6
	8.14	ND	71.0
	8.42	ND	43.1
	8.53	ND	46.7
	8.66	ND	24.4
	9.42	ND	13.2
	10.06	ND	5.5
	10.30	ND	6.8
	10.63	ND	5.4
	10.66	ND	3.8
	10.75	ND	6.3
	11.14	ND	5.5
250	7.35	ND	164.0
	8.03	ND	77.4
	8.44	ND	54.8
	8.72	ND	26.0
	9.23	ND	16.2
	9.36	ND	12.2

Table 2. Residual concentrations from batch coprecipitation/adsorption studies. (Continued)

Initial Ca^{++} Concentration, mg/l as CaCO$_3$	Final pH	Residual Soluble Concentration	
		Pb^{++}, mg/l	Ca^{++}, mg/l as CaCO$_3$
	9.74	ND	9.2
	10.02	ND	7.2
	10.35	ND	4.9
	10.42	ND	6.0
	10.71	ND	3.6
	11.08	ND	8.55
350	7.65	ND	181.0
	8.07	ND	118.0
	8.45	ND	40.3
	8.90	ND	27.1
	9.18	ND	19.6
	9.64	ND	14.0
	9.66	ND	14.6
	10.13	ND	10.7
	10.38	ND	5.5
	10.72	ND	5.5
	10.84	ND	7.1
	11.17	0.041	6.2

Pb$_i$ = 5.0 mg/l:

150	7.45	0.049	126.0
	8.05	0.033	65.5
	8.14	0.020	57.0
	8.64	ND	43.1
	8.69	ND	30.6
	9.05	ND	19.5
	9.41	ND	13.1
	9.55	ND	12.2
	9.91	ND	8.4
	10.46	0.069	6.7
	10.72	0.074	6.0
	10.84	0.099	6.8
250	7.56	0.071	128.0
	7.80	0.041	103.0
	8.16	ND	60.2
	8.56	ND	38.2
	8.86	ND	27.8
	9.06	ND	21.4
	9.80	ND	8.9
	10.08	ND	5.9
	10.28	ND	5.3
	10.43	0.030	4.2
	10.76	0.084	5.35
	11.02	0.070	6.9

Table 2. Residual concentrations from batch coprecipitation/adsorption studies. (Continued)

Initial Ca^{++} Concentration, mg/l as $CaCO_3$	Final pH	Residual Soluble Concentration	
		Pb^{++}, mg/l	Ca^{++}, mg/l as $CaCO_3$
350	7.58	0.121	137.0
	8.01	0.032	83.7
	8.11	0.080	62.5
	8.45	0.028	44.8
	9.19	0.034	15.6
	9.37	ND	9.5
	9.94	ND	4.7
	10.11	ND	7.3
	10.37	ND	7.2
	10.74	ND	2.8
	10.82	0.108	6.15
	11.11	0.143	3.4

NOTE: ND = Not Detected.

Table 3. Residual hardness at steady state conditions.

Series	Run No.	pH	Detention Time τ, min	Initial Hardness, mg/l as $CaCO_3$			Residual Hardness, mg/l as $CaCO_3$			%Hardness Removed		
				Ca^{++}	Mg^{++}	Total	Ca^{++}	Mg^{++}	Total	Ca^{++}	Mg^{+++}	Total
I	P-1	10.41	20.0	250	0	250	37	0	37	85.2	--	85.2
	P-2	10.35	30.0	250	0	250	58	0	58	76.8	--	76.8
	P-3	10.35	40.0	250	0	250	43	0	43	32.8	--	82.8
III	Pb-1	10.29	15.0	250	0	250	25	0	25	90.0	--	90.0
	Pb-2	10.30	20.0	250	0	250	48	0	48	80.8	--	80.8
	Pb-3	10.30	30.0	250	0	250	29	0	29	88.4	--	88.4
	Pb-4	10.30	20.0	250	0	250	31	0	31	87.6	--	87.6
V	Com-1	10.34	20.0	250	0	250	50	0	50	80.0	--	80.0
VII	Pb-5	10.79	15.0	250	65	315	58	41	99	76.8	36.9	68.6
	Pb-6	10.76	20.0	250	65	315	44	31	75	52.4	52.3	76.2
	Pb-7	10.80	30.0	250	65	315	53	19	72	78.8	70.8	77.1
X	M-1*	10.78	19.45	247.2	63.4	310.6	54.3	54.2	108.5	78.0	14.5	65.1
	M-2*	10.78	28.45	250.0	67.0	317.0	58.1	40.1	98.2	76.8	40.1	69.0
	M-3*	10.80	38.3	251.5	66.0	317.5	47.7	44.8	92.5	81.0	32.1	70.9

*Study performed by Pu-Hua Chen (27).

Table 4. Residual heavy metal concentration for the MSMPR study.

Series	Run No.	pH	Detention Time τ, min.	Initial Pb Concentration, mg/l	Lead Concentration, mg/l Range	Before Steady State	After Steady State	% Pb Removal After Steady State
III	Pb-1	10.29	15.0	1.22	ND	ND	ND	>99.2
	Pb-2	10.30	20.0	1.22	ND	ND	ND	>99.2
	Pb-3	10.30	30.0	1.22	ND	ND	ND	>99.2
	Pb-4	10.30	20.0	1.22	ND	ND	ND	>99.2
V	Com-1	10.34	20.0	1.0	0.003-0.008	0.003	0.008	99.8
VII	Pb-5	10.79	15.0	1.22	ND	ND	ND	>99.2
	Pb-6	10.76	20.0	1.22	ND	ND	ND	>99.2
	Pb-7	10.80	30.0	1.22	ND	ND	ND	>99.2

NOTE: ND = Not Detected.

Table 5. Alkalinity measurements from the MSMPR study.

Series	Run No.	Effluent pH	Alkalinity, mg/l as $CaCO_3$			
			OH^-	HCO_3^-	$CO_3^=$	Total
I	P-1	10.41	13.3	14.9	35.4	63.6
	P-2	10.35	11.2	10.2	21.0	42.4
	P-3	10.35	11.2	10.8	22.7	44.7
III	Pb-1	10.29	12.0	18.6	40.9	71.5
	Pb-2	10.30	12.2	10.4	23.1	45.7
	Pb-3	10.30	12.1	11.3	25.0	48.4
	Pb-4	10.30	12.7	10.6	24.6	47.9
V	Com-1	10.34	13.5	12.7	31.2	57.4
VII	Pb-5	10.79	39.2	6.6	47.2	93.0
	Pb-6	10.76	36.7	6.6	44.7	88.0
	Pb-7	10.80	40.2	6.5	47.9	94.6
X	M-1*	10.78	26.85	18.9	95.5	141.25
	M-2*	10.78	29.45	10.15	56.1	95.7
	M-3*	10.80	30.85	7.95	46.1	84.85

*Study performed by Pu-Hua Chen (27).

Table 6. Mean effluent conditions from the MSMPR study.

Series	Mean pH	Mean Alkalinity, mg/l as $CaCO_3$				Mean Hardness, mg/l as $CaCO_3$			Mean Residual Pb Concentration, mg/l
		OH^-	HCO_3^-	$CO_3^=$	Total	Ca^{++}	Mg^{++}	Total	
I	10.37	11.9	12.0	26.3	50.2	46.0	0.	46.0	---
III	10.30	12.3	12.7	28.4	53.4	33.3	0	33.3	ND
V	10.34	13.5	12.7	31.2	57.4	50.0	0	50.0	0.008
VII	10.79	38.7	6.6	46.6	91.9	51.7	30.3	82.0	ND
X*	10.79	29.05	12.35	65.9	107.25	53.35	46.4	99.75	---

*Study performed by Pu-Hua Chen (27).

NOTE: ND = Not Detected.

Table 7. Kinetic data for the individual experimental runs.

Series	Run No.	pH	τ, min.	G, µm/min.	n^o, no./ml-µm	B^o, no./ml-min.	L_D, µm	Correlation Coefficient, r	Approximate Calcite %
I	P-1	10.41	20.0	0.472	3200	1500	28.1	-0.999	0
	P-2	10.35	30.0	0.361	2300	860	33.0	-0.998	0
	P-3	10.35	40.0	0.300	1200	370	36.2	-0.994	0
III	Pb-1	10.29	15.0	0.377	17000	6400	17.0	-0.998	100
	Pb-2	10.30	20.0	0.243	21000	5100	14.6	-0.996	100
	Pb-3	10.30	20.0	0.268	15000	3900	16.1	-0.995	100
	Pb-4	10.30	30.0	0.163	22000	3500	14.7	-0.998	100
V	Com-1	10.34	20.0	0.306	7800	2400	18.4	-0.999	100
VII	Pb-5	10.79	15.0	0.279	39000	11000	12.6	-0.998	95
	Pb-6	10.76	20.0	0.214	41000	8800	12.8	-0.991	95
	Pb-7	10.80	30.0	0.193	17000	3300	17.4	-0.997	85
X	M-1*	10.78	19.45	0.297	24900	7410	17.3	-0.996	<1
	M-2*	10.78	28.45	0.262	7600	1990	22.3	-0.998	0
	M-3*	10.80	38.3	0.201	4760	955	23.1	-0.997	0

*Study performed by Pu-Hua Chen (27).

Table 8. Summary of the precipitation kinetics for each series of runs.

Series	Run Nos.	i	k_N	Average L_D, μm	Correlation Coefficient, r^2	95% Confidence Interval for i
I	P-1,2,3	3.02	1.55×10^4	32.4	0.975	+0.685
III	Pb-1,2,3,4	0.67	1.15×10^4	15.6	0.847	+0.297
VII	Pb-5,6,7	2.79	4.32×10^5	14.3	0.827	+1.849
X	M-1,2,3*	4.86	2.04×10^6	20.9	0.934	+1.84

*Study performed by Pu-Hua Chen (27).

AGGLOMERATION OF POTASSIUM SULFATE CRYSTALS IN AN MSMPR CRYSTALLIZER

J. Budz, A. G. Jones and J. W. Mullin ■ Department of Chemical and Biochemical Engineering, University College London, London WC1E 7JE, England.

Agglomeration of potassium sulfate crystals grown in aqueous suspensions in a continuous cooling MSMPR crystallizer has been studied using an optical technique. The degree of agglomeration has been assessed and found to be dependent on supersaturation and magma density and exhibit a maximum with crystal size. The effects of agglomeration on crystal growth and nucleation rates determined from the crystal size distributions are estimated.

The MSMPR crystallization technique is a convenient way to determine both growth and nucleation kinetics of crystallizing systems by analysis of the crystal size distribution (CSD). However, although it is possible in the laboratory to conform to the assumptions required by MSMPR theory (Randolph and Larson, [1]) there is one phenomenon which often occurs despite all efforts to exclude it, namely crystal agglomeration.

Agglomeration is an elusive process to quantify and has usually been ignored or assumed to be negligible in applications of population balance theory. However, it has been shown recently by Nývlt and Karel [2] and Karel and Nývlt [3] that even for such a simple and "well-behaved" system as potash alum crystallizing from aqueous solution, neglect of agglomeration during an MSMPR CSD analysis can give rise to errors in the estimated growth and nucleation rates as large as 45% and 70% respectively.

When considering agglomeration processes in crystallizing suspensions at least two separate cases can be distinguished. In the first agglomeration mainly occurs in the sub-micron size range and there are no well developed single crystals (Shah, [4]; Matusevich and Shabaline, [5]; Sakamoto et al, [6]). In this case, Shah [4] and Sakamoto et al [6] claim that the rate of agglomeration is greatest (i) when the rate of crystallization is high and (ii) when the size of particles in the solution is small. These conditions are met in most precipitation processes. The second case is when crystallization occurs at a low supersaturation level, for example, crystallization by cooling of a soluble inorganic salt in a fluidised bed or MSMPR type crystallizer. Under such hydrodynamic conditions the first particles formed are usually well developed single crystal fragments and their subsequent agglomeration occurs over the whole range of sizes, as observed by Karel and Nývlt [3] for potash-alum, Shah [4] for nickel-ammonium sulfate and in this present study of potassium sulfate.

The mechanism of agglomeration is very complicated and depends on several parameters. Matusevich and Shabaline [5] concluded that degree of agglomeration decreases with increasing intensity of agitation, mean crystal size as well as solids concentration (magma density) while Mullin [7] and Misra and White [8] found that agglomeration increases with supersaturation and reaches a maximum at a certain level.

Three main techniques are available for estimating the degree of agglomeration in crystallizing systems. The first two techniques do not determine agglomeration directly but infer it from a curve-fit to a chosen model. The first computational method to derive growth, nucleation and agglomeration kinetics from steady-state CSD was proposed by Liao and Hulburt (9). Tavare et al (10) used this method to deduce the kinetics of agglomeration of nickel ammonium sulfate crystals precipitated in an MSMPR crystallizer. The methods of Nývlt and Karel (2) and Karel and Nývlt (3) involve decomposition of the CSD and are based on the assumption that the true CSD (without agglomeration) corresponds to a given theoretical relation. In all these cases, the experimental data fit the proposed model well, but their interpretation depends on the choice of model.

In the present work, however, the aim was to determine agglomeration directly in order to determine the true CSD. Thus a third, albeit somewhat tedious (but most reliable) visual evaluation technique (Nývlt and Karel, 2; Karel and Nývlt, 3; Shah, 4) was adopted and applied to potassium sulfate crystallized by cooling over a range of supersaturation, magma density and temperature. Subsequently, the degree of agglomeration determined was used to refine population density CSD data obtained by sieve analysis and its effect on apparent crystal growth and nucleation rates estimated.

THEORY

The population density of particles, n, is defined by:

$$n = \lim_{\Delta L \to 0} \left(\frac{\Delta N}{\Delta L} \right) \quad (1)$$

Thus for a particular size range, ΔL, the population density, n_i, is given by:

$$n_i = \frac{m_i M_T}{\alpha \rho_C L_i^3 \Delta L \sum_i m_i} \quad (2)$$

where m_i is a mass of particles of mean size L_i, M_T is magma density and the subscript i refers to the i-th size range.

Since during sieving crystals of a given size are distributed between sieves both as discrete crystals and as agglomerates, then the sieve analysis obtained must be corrected for agglomeration to obtain the true distribution.

For regular crystals the characteristic size obtained by sieving is the crystal width, that is the second largest dimension. Geometrical considerations lead to the conclusion that for an agglomerate comprising randomly stuck equi-size particles the agglomerate size is approximately equal to the constituent particle length. Thus the characteristic size of the consitituent crystals of the agglomerates retained on sieve i is equal to L_i/K_i where K_i is the crystal length to width, or aspect, ratio.

The number-based degree of agglomeration for the i-th size range is defined as:

$$P_i = \left(\frac{N_a}{N_a + N_c} \right)_i \quad (3)$$

where N_a = number of agglomerates, and N_c = number of discrete crystals.

Assuming that all crystals in a given agglomerate are the same size then the mass of crystals on the i-th sieve, m_i', is given by:

$$m_i' = N_{ci} \alpha \rho_C L_i^3 + k N_{ai} \alpha \rho_C (L_i/K_i)^3 \quad (4)$$

where k = number of crystals involved in one agglomerate and (L_i/K_i) = size of individual crystals creating an agglomerate.

The mass of all crystals of size L_i can be calculated from the relation:

$$m_i = m_i' - \Delta M_{Ai} + \Delta M_{Bj} \quad (5)$$

where ΔM_{Ai} is the mass of agglomerates retained on the i-th sieve comprising crystals smaller than L_i:

$$\Delta M_{Ai} = k N_{ai} \alpha \rho_C (L_i/K_i)^3 \quad (6)$$

and ΔM_B is the mass of crystals of size L_i constituting agglomerates, which are retained on a larger sieve-j:

$$\Delta M_{Bj} = k N_{aj} \alpha \rho_C (L_j/K_j)^3 \quad (7)$$

where $(L_j/K_j) = L_i$.

Values of ΔM_A and ΔM_B are related and for practical purposes ΔM_{Aj} is subtracted from the mass of the j-th sieve cut and distributed among smaller size fractions corresponding to the size (L_j/K_j).

Defining the mass-based degree of agglomeration, $\emptyset_i$, as

$$\emptyset_i = m_i/m_i^c \qquad (8)$$

the "real" or refined population density, n_i, can now be calculated from the equation:

$$n_i = \frac{m_i M_T}{\alpha \rho_C L_i^3 \Delta L \Sigma m_i^c} = \emptyset_i n_i^c \qquad (9)$$

The calculation procedure is as follows. Firstly the number based degree of agglomeration, P_i, is determined (e.g. visually). Then, knowing the value of m_i^c from a sieve analysis, values of N_{ai} and N_{ci} are calculated (Equations 3 and 4), then m_i and $\emptyset_i$ (Equations 5 and 8) are determined and finally the refined value of population density, n_i, (Equation 9) is evaluated.

EXPERIMENTAL

All experiments were performed using aqueous potassium sulfate solutions in a small laboratory-scale MSMPR crystallizer (300 mL capacity, stirrer speed 1000 rpm, operated at steady-state) described in detail elsewhere (Budz et al, 11). After sieve analysis of the dried crystals the degree of agglomeration was determined visually by counting the number of agglomerates, N_a, and of all particles, ($N_a + N_c$), under an optical microscope (Olympus BH). At least 150 particles were counted for each sieve cut. Altogether 18 experiments were performed at the temperature of 20°C (10 runs) and 50°C (8 runs) within a range of operational parameters of supersaturation σ (0.03 - 0.12); magma density M_T (2 - 18 kg/m³) and residence time τ (17 - 59 mins). Relative supersaturation, σ, was defined as:

$$\sigma = \frac{w - w_{eq}}{w_{eq}} \qquad (10)$$

where w and w_{eq} are prevailing and equilibrium concentrations respectively.

RESULTS AND DISCUSSION

Degree of Agglomeration

Crystals building up the agglomerates were mostly randomly orientated and only a few regularly shaped doublets and triplets were observed (see Figure 1). The average number of crystals involved in one agglomerate, k, was estimated to be 3. For the crystals under consideration the length to width ratio $K = 2\alpha$, where α is volumetric shape factor and is size-dependent (Mullin and Gaska, 12; Budz et al, 13).

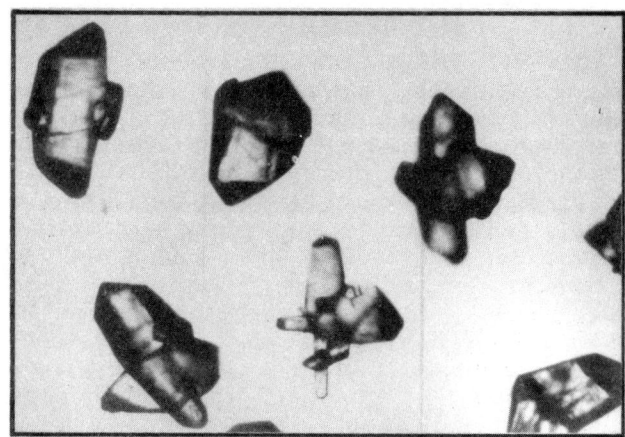
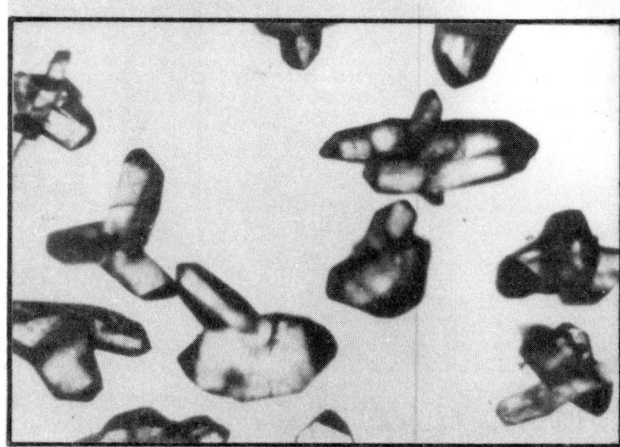

Figure 1. Potassium sulphate crystal agglomerates.

The experimentally determined degree of agglomeration as a function of size (Figure 2) was best correlated with supersaturation, σ, and magma density, M_T, by a log-normal distribution of the form:

$$P(L) = P_O \exp(-A \log^2(L/L_O)) \qquad (11)$$

where
$$A = 1.7$$
$$L_O = 930\, \sigma^{0.64} \qquad (12)$$

and
$$P_O = 0.34\, (M_T \sigma)^{0.36} \qquad (13)$$

No effect of temperature was found over the range 10-50°C.

The error in the correlation shown in Figure 2 does not exceed ± 20% for all runs. This is more than satisfactory bearing in mind the random nature of the process under investigation.

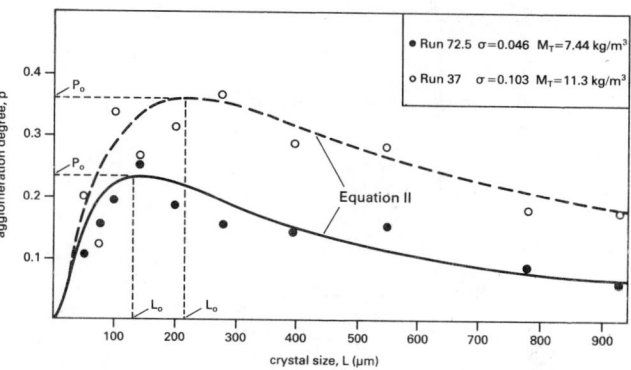

Figure 3. Degree of agglomeration, (K_2SO_4 crystals).

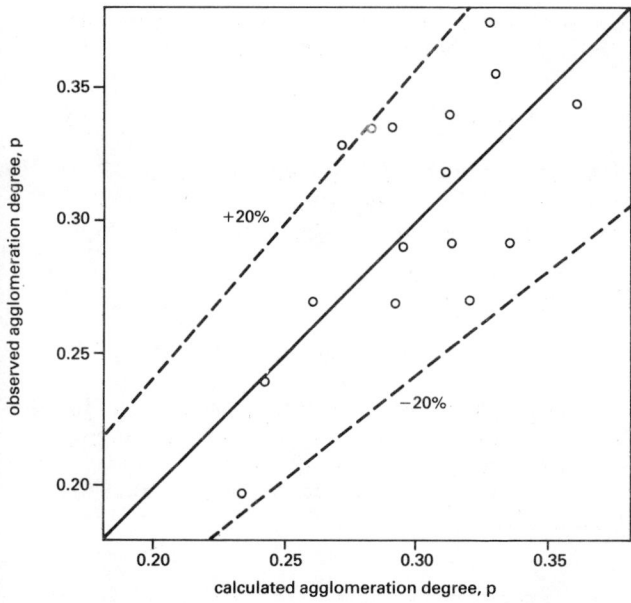

Figure 2. Degree of agglomeration correlation, (Equation II).

The good fit of this correlation to the experimental results is also shown for two chosen runs in Figure 3. Interestingly, a maximum is observed on the P-L curve and this can be tentatively explained as follows.

In the MSMPR crystallizer, there are two competitive processes influencing the final degree of agglomeration: (i) agglomeration of crystals and (ii) breakage of agglomerates already formed. The rate of the former processes increases with increasing size of crystals; crystals older and thus larger having more chances to collide and agglomerate. Large particles also undergo more breakage as increased size causes more energetic contacts between crystals and internal parts of a crystallizer. The retarding process of breakage finally increases faster with size than agglomeration itself. Superposition of these two processes results in a maximum as recorded experimentally (Figure 3). The size at which the P-L curve reaches the maximum, L_O, is proportional to supersaturation (Equation 12) which again can be explained by the fact that rate of agglomeration increases with supersaturation whereas the energy of collisions is independent of it. As a result, the recorded peak is shifted towards larger sizes for higher levels of supersaturation. It is also possible that some agglomerates arise from a growth mechanism.

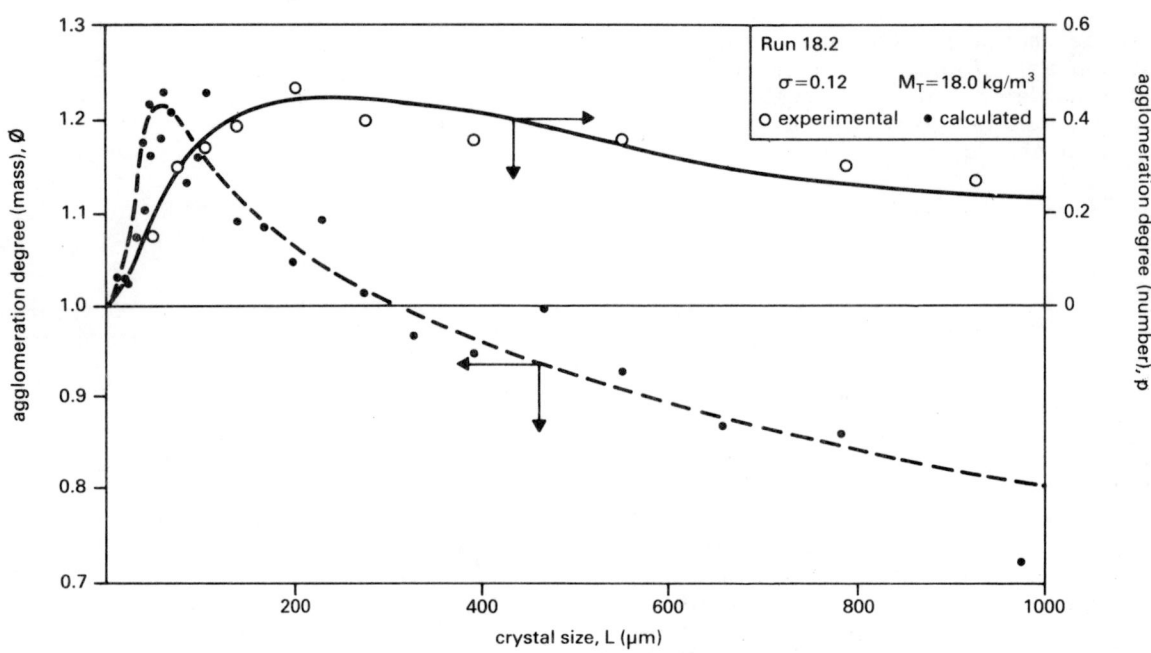

Figure 4. Comparison between number (solid line) and mass (dotted line) based on degree of agglomeration.

The degree of agglomeration on a mass basis, which reflects the difference between mass distributions of crystals calculated with and without agglomeration, for one of the experiments (run 18.2) is presented in Figure 4. It can be seen that agglomeration simply shifts the mass distribution towards larger sizes around a certain size larger than the size L_O but apparently related to it. The difference between the real and appearent (i.e. measured) distributions reaches maximum at about 50 μm and then decreases sharply down to negligible level.

Growth and Nucleation Rates

It is of interest to analyse how agglomeration can affect calculations of crystal growth and nucleation rates in an MSMPR crystallizer from population density data. In order to calculate the strongly size-dependent growth rate of potassium sulfate crystals, the equation of Sikdar (14) was used:

$$G(L) = \frac{N(L)}{n(L)\,\tau} \qquad (14)$$

where G(L) = linear growth rate along the characteristic crystal dimension N(L) = cumulative number oversize, n(L) = population density, τ = mean residence time.

An excellent fit of the experimental data to the size-dependent growth model for potassium sulfate proposed by White et al. (15) was obtained.

From calculations without taking agglomeration into account for an example run (18.2), the crystal growth rate is given by:

$$G' = 4.267 \times 10^{-10} (1+2L)^{2/3} \quad \text{m/s} \qquad (15)$$

with coefficient of correlation = 0.936 (L is expressed in microns). Whereas for calculations corrected for agglomeration:

$$G = 4.057 \times 10^{-10} (1+2L)^{2/3} \quad \text{m/s} \qquad (16)$$

with coefficient of correlation = 0.946.

Thus correcting for agglomeration in the growth rate calculation gives a difference compared with unrefined data of about −10% and a slightly better fit to the growth model (Figure 5).

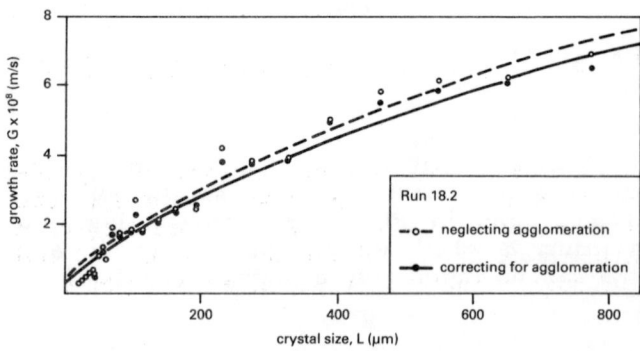

Figure 5. Effect of neglecting agglomeration on calculated growth rates.

All such calculations become uncertain for very small crystal size range where of course other size-dependent events can occur, the most probable being secondary nucleation by collisions and growth dispersion. It can be seen from Figure 5 that the calculated growth rates for crystals of size smaller than about 50 μm are lower than those predicted by the general size-dependent growth model.

Thus conventional estimation of nucleation rate based on extrapolation of the population density-size curve to zero size is not appropriate here. The method of calculation of so-called "standard nucleation rate" described by Jančić and Grootscholten (16) offers an alternative approach. In this method nucleation at zero crystal size, $B°(=n°G°)$, is replaced by the convective number rate at size L^*, B^* ($=n^*G^*$), with $L^* = 50$ μm being chosen in the present case. For the chosen run (18.2) the nucleation rate uncorrected for agglomeration is given by:

$$B^{*\prime} = 157 \times 10^3 \text{ particle/s m}^3 \quad (17)$$

whereas corrected for agglomeration

$$B^* = 202 \times 10^3 \text{ particle/s m}^3 \quad (18)$$

Thus nucleation rates corrected for agglomeration are about + 30% higher. This error (obtained at comparatively low magma densities up to 18 kg/m³) is not very large, but might be expected to be more significant at higher values of magma density, as for example those used in the work of Randolph and Rajagopal (17) (M_T up to 28 kg/m³) or Rosen and Hulburt (18) (M_T up to 108 kg/m³) for the same substance.

CONCLUSIONS

Visual determination of the degree of agglomeration has proved to be an useful tool in the quantitative assessment of crystal agglomeration and its effect on the estimation of growth and nucleation rates. The effective degree of agglomeration of potassium sulfate crystals in a continuous MSMPR cooling crystallizer is a function of crystal size and increases with magma density and supersaturation. It appears to be a result of a competition between the formation of agglomerates and their subsequent breakage due to collisions. A maximum in the degree of agglomeration at about 50 μm supports that view.

Neglect of the effect of agglomeration may result in large errors to occur in the determination of kinetic data from conventional steady-state CSD analysis. In this paper deviations to growth and nucleation rates of -10% and +30% respectively are reported, and substantially higher errors may be expected in the more concentrated suspensions encountered in practice.

Nomenclature

A — constant (Equation 11)
B — nucleation rate, particles/s m³
G — growth rate, m/s
k — number of crystals in an agglomerate,-
K — length to width ratio of crystals,-
L — crystal size, m, μm
M_T — magma density, kg/m³
N — cumulative number oversize, particles/m³
m — mass, kg
n — population density, particles/m m³
P — number-based degree of agglomeration,-
w — concentration, kg/kg free water

Superscript

′ — un-corrected for agglomeration

Subscripts

a — agglomerate
eq — equilibrium
c — crystal
i,j — size increments

Greek letters

α — volumetric shape factor, -
$ρ_c$ — crystal density, kg/m³
Δ — difference
Σ — sum
∅ — mass-based degree of agglomeration,-
σ — relative supersaturation,-
τ — residence time, s

REFERENCES

1. Randolph A.D., Larson M.A., Theory of Particulate Processes, Academic Press, New York. 1971.

2. Nývlt J., Karel M., Crystal Res. & Technol. 20, 173, (1985).

3. Karel M., Nývlt J., ibid, 447, (1985).

4. Shah M.B., "Crystallization kinetics and agglomeration of nickel ammonium sulfate in a continuous crystallizer", Ph.D. Thesis, University of London, (1980).

5. Matusevich L.N., Shabaline K.N., Zh. Prikl. Khim., $\underline{25}$, 1157, (1952).

6. Sakamoto K., Kanehara M., Matsushita K., J. Chem. Eng. Japan, $\underline{35}$, 481, (1971).

7. Mullin J.W., Crystallization. 2nd Ed., Butterworths, London, 1972.

8. Misra C., White E.T., J. Crystal Growth, $\underline{8}$, 172-178, (1971).

9. Liao P.F., Hulburt H.M., "Agglomeration Process in Suspension Crystallization". Paper presented at Annual Meeting of AIChE, Chicago, December, (1976).

10. Tavare, N.S., Shah, M.B., Garside, J., Powder Technol., $\underline{44}$, 13-18, (1985).

11. Budz J., Jones A.G., Mullin J.W., J. Chem Tech. Biotechnol., $\underline{36}$, 153-161, (1986).

12. Mullin J.W., Gaska C., J. Chem. Eng. Data, $\underline{18}$, 217, (1973).

13. Budz J., Jones A.G., Mullin J.W., "On Shape-Size Dependency of Potassium Sulphate Crystals", Ind. Eng. Chem. Res. In press.

14. Sikdar S.K., Ind. Eng. Chem. Fundam., $\underline{16}$, 390, (1977).

15. White E.T., Bendig L.L., Larson M.A., AIChE Sympos. Ser. No. 153, $\underline{17}$, 41, (1976).

16. Jančić S.J., Grootscholten P.A.M., Industrial Crystallization, Delft University Press, Delft, Holland, 1984.

17. Randolph A.D., Rajagopal K., Ind. Eng. Chem. Fundam., $\underline{9}$, 165, (1970).

18. Rosen H.N., Hulbert H.M., Chem. Eng. Prog. Sympos. Ser., $\underline{67}$, 110, 18, (1971).

BIMODAL CSD BARITE DUE TO AGGLOMERATION IN AN MSMPR CRYSTALLIZER

James R. Beckman and Robert W. Farmer ■ Department of Chemical and Bio Engineering, Arizona State University, Tempe, AZ 85287

Crystallization processes in which agglomeration is an important particle enlargement mechanism are frequently encountered industrially; however, appropriate models have not been sufficiently developed. Experimental precipitation of barium sulfate (barite) in a clear-liquor-advance MSMPR crystallizer has shown that the solid product generally displays a bimodal CSD, which was attributed to strong agglomeration. A population balance including terms representing the agglomerative mechanism was used to interpret the interaction of primary and agglomerative growth. Values of an empirical agglomeration efficiency function indicate that barite agglomeration is enhanced by high precipitation rates.

It has been observed in some crystallization processes of industrial importance that crystal agglomeration contributes significantly to particle growth (Maruscak et al. (1), Baker and Bergougnou (2), Halfon and Kaliaguine (3), Sarig et al. (4). Generally a collision-cementation mechanism is posed as leading to agglomerate formation. First, random binary crystal collisions driven by turbulent motion within the crystallizer must result in temporary adhesion of the collided particles. If the collided crystals remain in contact for a sufficient length of time, intergrowth of new solid phase by deposition solute material will cement the particles together, yielding an agglomerate.

In this study, the precipitation of barium sulfate (barite) was examined under conditions of high precipitation rate in which agglomeration was a significant particle enlargement mechanism. Barite product CDS obtained from an MSMPR crystallizer having clear-liquor-advance exhibited a characteristic bimodal form. A population balance incorporating terms for primary and agglomerative growth was applied to model these distributions, and interpret the interactions between the two growth modes.

Clearly, not all binary crystal collisions will lead to successful formation of an agglomerate. To quantify this complicated phenomena, it is necessary to define a suitable agglomeration efficiency (i.e. collision frequency function). A hypothesis of this study is that appropriate efficiency function forms may be constructed by phenomenological reasoning based on several important physical factors:

1. Relative particle size - Crystals of similar size tend to deflect each other, and if grossly different in size the smaller would be swept past the larger. Between these extremes a "kinematic capture" mechanism is active, resulting in favorable adhesion conditions.

2. Absolute agglomerate size - Larger incipient agglomerates would be subjected to more severe fluid-mechanical stresses, and would be more likely to be ripped apart prior to intergrowth.

3. Agitation intensity (i.e. impellor speed) - Favorable energetaics of collision must be balanced against the level of mechanical stresses on collided/adhered crystal pairs.

4. Solute supersaturation - Higher local supersaturation levels (even if undetectable) would result in more rapid intergrowth, reducing the contact time necessary for agglomerate formation.

In previous modeling investigations of agglomeration, factors (3) and (4) above have been acknowledged (Hulbert and Katz (5), Baker and Bergougnou (2), Halfon and Kaliguine (3)). However, size-dependencies such as are implied by factors (1) and (2) have been either neglected or treated rather simplistically.

Many investigators have studied the nucleation and growth of barium sulfate. Much of the previous work on this topic has been limited to growth at very low ionic concentrations (i.e. $[Ba^{++}] < 10^{-4}$ M). Under such conditions the precipitation rate is very low, and there is a substantial induction period before nucleation is initiated (Nielsen (6), (7); Gunn and Murth (8)).

In a seeded crystallizer at high precipitation rates it has been found that nucleation is essentially instantaneous, and that primary crystal growth follows second-order kinetics with respect to supersaturation (Nancollas and Purdie (9), Rizkalla (10)). Nancollas and Liu (11) and Liu et al. (12) examined the morphology of precipitated barite as a function of ion concentration, agitation intensity, and the presence of foreign metal ions. At barium ion concentrations greater than .001 M it was observed, by scanning electron microscopy, that barium sulfate grew in the form of orthogonal, planar, dendritic crystals. Competition between surface reaction and diffusional limitations were recognized as causative factors of crystal habit. Early observations indicated that agglomeration of dendritic barite crystals occurred at sufficiently high precipitation rates (Fischer (13), (14)).

AGGLOMERATION/GROWTH MODEL

The population density function ($n(L)$) for agglomerating particulate systems can display a characteristic inflection point (Kubota and Mullin (15)). For strongly agglomerating systems, $n(L)$ may even have a bimodal form (Halfon and Kaliaguine (3), (16)). In effect, it appears that greater numbers of crystals are introduced into larger size regions than would appear by purely primary growth of smaller crystals. To mathematically describe this characteristic CSD shape additional secondary growth terms must be introduced into the governing population balance.

Secondary growth via agglomeration can be represented by integral terms associated with the appearance or removal of particles with respect to a given size. The integrand may be expressed as the number of successful binary collisions which can lead to birth of a particle of size L (Hulbert and Katz (5)):

$$B = V \int_0^L \beta_b(L',L'')n(L')n(L'')dL' \quad (1)$$

Similarly, the death of a particle of size L may be expressed as

$$B = V \frac{1}{2} \int_0^\infty \beta_b(L',L)n(L')n(L)dL' \quad (2)$$

Only two-particle collisions are considered in equations (1) and (2); therefore, this simple collision model represents an analogue of the classical Boltzmann equation in the kinetic theory of dilute gases. An additional constraint on L' and L'' is that their relative sizes are dictated by additive volumes:

$$L^3 = L'^3 + L''^3 \quad (3)$$

Furthermore, numerical calculations of the collision integrals assign L'' > L' as the relative sizes of colliding crystals; thus L'' may be termed the "capturing" particle, and the smaller L' the "captured" particle.

A collision efficiency factor, $\beta(L', L'')$, is employed that is essentially the probability that two colliding crystals will adhere and yield a stable agglomerate. The $\frac{1}{2}$ prefactor in equation (2) insures that symmetric collisions are counted only once.

The steady state population balance for an MSMPR crystallizer, including agglomerative growth terms, is written as:

$$B - D = \text{growth} + \text{outage} \quad (4)$$

$$\int_0^L \beta_b n(L')n(L'')dL' - \frac{1}{2}\int_0^\infty \beta_d n(L')n(L'')dL'$$
$$= G\frac{dn}{dL} + \frac{n}{\tau} \quad (5)$$

where: $\tau = V/Q_p$

Various empirical and theoretical forms of the agglomeration efficiency, β, have been proposed (Schumann (17), Thompson (18), Drake (19), Bapat et al. (20), Ramabhadran et al. (21), Halfon and Kaliaguine (16)). Several such kernel functions were compared with three models generated in this study. Table 1 lists the kernel functions which were evaluated.

Table 1
Agglomeration Kernel Functions Evaluated

Model I:
$$\beta(L',L'') = k \frac{L' L''}{(L'^3 + L''^3)^\gamma} (e^{-\lambda L'})$$

Model II:
$$\beta(L',L'') = k \frac{L' L''^2}{(L'^3 + L''^3)^\gamma} (e^{-\lambda L'})$$

Model III:
$$\beta(L',L'') = k \frac{L''^2}{(L'^3 + L''^3)^\gamma} (e^{-\lambda L'})$$

Schumann (17):
$$\beta(L',L'') = k \frac{(L' + L'')(L' L'')^{1/2}}{(L'^3 + L''^3)^{3/2}}$$

Thompson (18):
$$\beta(L',L'') = k \frac{(L''^3 - L'^3)^2}{(L'^3 + L''^3)}$$

Bapat (20):
$$\beta(L',L'') = k \frac{(L' + L'')^2 (L'^{2/3} + L''^{2/3})^{1/2}}{\exp\left[\lambda \left(\frac{L' L''}{L' + L''}\right)^4\right]}$$

Constant Prefactor:
$$\beta(L',L'') = k$$

Exponential Decay:
$$\beta(L',L'') = k \exp\left[-\lambda (L'^3 + L''^3)^{1/3}\right]$$

DESCRIPTION OF EXPERIMENTS

Barite precipitation was performed in a Mixed-Suspension, Mixed-Product-Removal (MSMPR) crystallizer with clear-liquor-advance (CLA). The variables of interest were the barite production rate, and the agitation intensity, as measured by the impellor speed. While this is only a small subset of potential variables, these two factors encompass both supersaturation and hydrodynamic effects on barite growth. The experimental conditions are summarized in Table 2, which appears in the discussion of results.

A baffled, 2.4 liter crystallizer was fabricated from 10.2 cm (4 inch) I.D. acrylic tubing. Several wall ports were provided to permit sampling of the crystallizer magma from within the vessel. A custom-built syringe solenoid valve, actuated by an electronic repeat-cycle timer, permitted intermittent product withdrawal. Two-bladed acrylic paddles (7.5 cm x 2.5 cm) having rounded edges and polished surfaces were employed to provide good mixing while keeping secondary nucleation effects at a low level. Three such impellors were attached to a single stainless steel shaft.

The volumetric product flow rate was set to provide a 55-minute solids residence time for all MSMPR-CLA experiments. A second, larger volume overflow stream was routed to a conical glass clarifier, 1.2 liters in volume, and having 15.2 cm (6 inch) top diameter. Concentrated solids from the bottom of the clarifier were returned to the crystallizer by an air-lift pump, to avoid crystal attrition. This feature permitted the solids residence time to be considerably longer than liquid residence time. Typical fines loss via the clarifier overflow stream were between 1 to 3 weight percent of the barite product rate, mostly in the sub-3 micron size range.

For barite precipitation, feeds to the crystallizer consisted of anion and complementary cation aqueous solutions. Anhydrous, reagent grade citric acid (2.0 g/L) was added to the 0.25 M sodium sulfate (Na_2SO_4) stock solution, as an anti-scaling agent. Preliminary trials indicated that the presence of citrate anion had no detectable effect on the barite morphology or CSD in the growth rate regime of these experiments. The concentration of the barium cation feed (C_o) was varied to reflect the specified product rate level, and was prepared from reagent grade $BaCl_2 \cdot 2H_2O$ crystals. For all crystallization runs, the sulfate feed was diluted with a deionized water stream to maintain the $Ba^{++} SO_4^{=}$ ion ratio at 0.5, and the liquid residence time at 17 minutes.

To verify steady state operation and homogeneous mixing, the barite solids CSD and concentration were sampled both in the product stream and directly from the crystallizer magma. CSD determinations were performed on an 80XY electronic size analyzer from Particle Data Inc., using a 150 μm orifice and doubly filtered (0.45 μm) crystallizer liquor as electrolyte. In general, steady state conditions were achieved in six to eight residence times.

BARITE AGGLOMERATION MECHANISM

The barite product CSD data shown in Figure 1 illustrates the characteristic bimodal shape. The distributions shown were obtained from the crystallizer product stream and two locations within the vessel over a time span exceeding one solids residence time. The precision between these CSDs verifies that steady state generation and homogeneous mixing was maintained in this apparatus, even at the low impellor speeds used.

Inspection of these bimodal distributions suggests that barite growth behavior can be envisioned, and modeled, as comprising three regions with respect to size. In the sub-5 micron range, fines destruction via agglomeration is the predominant factor. At the largest sizes, the linearity of the differential population distribution, $n(L)$, can be represented by simple primary growth. Between these two size regimes, primary and agglomerative growth modes are competitive, yielding the bimodal shape which was observed in all the barite crystallization runs.

To verify this presumed mode of barite growth, product crystals were examined by scanning electron microscopy (SEM). This procedure revealed the nature of both single crystal and agglomerate morphology. Single crystals were planar in shape, consisting of dendrites grown along orthogonal axes. As particle size increased, the predominant morphology was agglomerates of these smaller crystals. Typical agglomerates consisted of a number of planar dendritic crystallites, juxtaposed at random angles, and inter-grown. Given the strong preference of $BaSO_4$, under high local supersaturations, to follow orthogonal growth patterns, the randomness of the observed dendritic plane angles could not be explained by twinning or dislocation growth from a single nuclei. It was apparent that particle growth occurred via the dual mechanism random collision followed by inter-growth, and steady plating growth of the cemented, planar crystallites.

Crystal breakage or attrition could also have accounted for the bimodal form of the barite CSD. It was found that these effects were minor based upon the results of suspension aging tests which consisted of extended mixing of the crystallizer magma with the feed and product streams shut off. Initially the agitation was held at the level at which the crystals were formed, followed by a period with increased or decreased impellor speed.

Figure 1 presents the results for one of the aging test sequences, in which impellor speed was increased. No significant change occurred during the initial period at unchanged agitation. When the impellor speed was increased there was a detectable increase in the number density of particles in the 3 - 12 μm range, and the CSD then stabilized following the first hour of more intense mixing.

It is proposed that this CSD shift was the result of breakage among the largest, most fragile, and least thoroughly inter-grown agglomerates; note that the only significant decrease in $n(L)$ occurs at the larger sizes present. The fact that the increase in $n(L)$ at smaller sizes did not continue indefinitely indicates that there were only a limited number of crystals within the agglomerated product population that could not withstand an increase in fluid-mechanical stresses. It was found that the barite CSD remained unchanged under constant agitation, meaning that particle breakage was negligible. Furthermore, there was not continued particle enlargement by agglomeration in the absence of fresh solute feed. This demonstrated that inter-growth was essential for the creation of barite agglomerates.

AGGLOMERATION EFFICIENCY MODEL EVALUATION

A suitable agglomeration efficiency model should result in birth and death integral values which follow appropriate qualitative trends. The contribution of agglomerative death is expected to be highest for smaller sized crystals and decrease nearly monotonically as size increases. The birth integral should exhibit a preferred size region for agglomerative birth.

As trial and error model screening progressed it was noticed that certain mathematical characteristics, such as a decaying exponential factor, produced favorable results. Also, it was consistently found that models containing one or two constants were not sufficiently flexible, that is, they did not have adequate parametric sensitivity, to permit close prediction of barite product CSD. For this reason, it was necessary to resort to three-constant empirical models for β_b and β_d.

Three model forms emerged as being well-suited to describe the size-dependent nature of barite agglomeration efficiency:

$$\text{model I)} \quad \beta(L',L'') = k \frac{L''L' e^{-\lambda L'}}{(L'^3 + L''^3)} \gamma \quad (6)$$

$$\text{model II)} \quad \beta(L',L'') = k \frac{L''^2 L' e^{-\lambda L'}}{(L'^3 + L''^3)} \gamma \quad (7)$$

$$\text{model III)} \quad \beta(L',L'') = k \frac{L''^2 e^{-\lambda L'}}{(L'^3 + L''^3)} \gamma \quad (8)$$

The alternative models listed in Table 1 are distinguished by lack of an appropriate physical basis, that is, they are either purely empirical, or they apply to physical conditions differing from crystal agglomeration systems.

In order to obtain an optimal set of model constants for a given MSMPR-CLA data set, the governing population balance (Equation (5)) was numerically integrated using a fourth-order Runge-Kutta method. The initial condition for the integration was to set $n(L_o) = n(3 \mu m)$, as obtained directly from CSD data. The primary growth rate, G, was obtained for each run by setting the CSD slope in the large-size regime equal to $-1/G\tau$. Estimates of the agglomeration birth and death integral values were calculated prior to the integration, based on the experimental n(L) data values and the current set of agglomeration efficiency model constants. Each integral was replaced with numerical summations over discrete size increments that matched the 80XY size analyzer channel increments.

A Hooke and Jeeves numerical search routine (Hooke and Jeeves (22), Kuester and Mize (23)) was employed to obtain a constant set based on a logarithmic sum-of-squares objective function:

$$SSQ = \sum_{i=1}^{P} \left(\ln n(L)_{data} - \ln n(L)_{calc} \right)_i^2 \quad (9)$$

where P represents the 80XY size analyzer data channel, corresponding to the largest detected crystal size. This measure of total model deviation is more appropriate than the sum of squared absolute deviations for data spanning several orders of magnitude. For each search iteration, an updated model constant set was supplied by the Hooke and Jeeves routine, new values of the agglomeration integral values were compiled, equation (5) was again integrated, and a new SSQ computed. This procedure continued until computed u(L) values converged with a constant set corresponding to a minimal SSQ.

Based on these numerical evaluations, five of the nine models listed in Table 1 were judged to be competitive. The final SSQ results for these cases are presented in Table 3. In all cases, the optimal constant set obtained from the numerical search, and the final SSQ values, were not significantly changed by posing widely differing initial guesses for the constants.

The agglomeration efficiency models proposed for non-crystallization systems could not adequately represent the size-dependence of agglomerative barite growth. These models exhibited final SSQ values generally a factor of three to ten higher than their counterparts obtained using models I, II, III, or two-constant empirical forms for β.

It is possible to statistically compare the robustness of the five candidate agglomeration efficiency models by establishing confidence intervals on differences in mean SSQ values. Model II had the smallest mean SSQ, so was used as the reference value for the SSQ differences. Because of relatively high variances, the discriminations can only be supported at the 80% level of significance. The computed confidence intervals shown in Table 2 are entirely or predominantly positive for the inverse power law, exponential decay, and model I kernel forms, meaning that these are less robust than model II. On a statistical basis, neither model II nor model III are favored. However, intuitive physical considerations favor model II, since this model predicts $\beta = 0$ at $L' = 0$, while model III predicts $\beta = \infty$.

Based on this analysis, model II was selected as a suitable representation of agglomeration efficiency. Taken separately each factor in this model may be connected to a physical interpretation of the agglomeration phenomenon. It seems reasonable that the likelihood of successful agglomeration should increase proportionally to capturing particle surface area (L''^2). This implies that agglomeration efficiency improves as the projected area swept by a crystal increases. The functionality of the captured particle diameter is a Gamma distribution ($L'\exp(-\lambda L')$) indicating that there is an "optimal" particle size for capture. This is consistent with the presumed mechanims of crystal contact and intergrowth. Finally, the inverse proportionality of final agglomerate volume ($1/(L'^3 + L''^3)$) may be related to the particle momentum, that is, its inertia. A smaller incipient agglomerate would more closely follow turbulent fluid motion, reducing the tendency to be broken apart by shear stresses. Thus smaller agglomerates have a better chance to become intergrown.

DISCUSSION OF RESULTS

Table 2 provides a summary of the experimental conditions for the eleven MSMPR/CLA crystallizer runs, along with the primary growth and nucleation parameters derived from the product CSD. As mentioned previously, the primary growth rate was estimated from the slope of the linear $n(L)$ curve in the larger-size regime. The total nucleation rate, B^o, was obtained by correcting this primary growth rate for fines destruction in the sub-5 micron size range (Randolph and Larson (24)):

$$B^o = \frac{G\, n^o}{R} \quad (10)$$

where R is defined as the slope ratio;

$$R = \left.\frac{dn(L)}{dL}\right|_{L<5\ \mu m} - \frac{1}{G\tau}. \quad (11)$$

It is noteworthy that the relative trends with respect to agitation intensity in average particle size, primary growth rate, and nucleation rate differ depending on product rate. Referring to Table 3, at 0.5 g/min product rate the average size and growth rate decrease, and the nucleation rate increases, with higher impellor speed. But at a 1.0 g/min product rate these relative trends are reversed, and it can be shown that this is a result of agglomerative growth. This observation also suggests that agglomeration is influenced both by shifts in local supersaturations (i.e. caused by changes in product rate) and in suspension agitation, but to different degrees.

Figure 2 shows the complete product size distributions observed across the range of product rates tested. Clearly, barite product rate has a large effect on crystal size, with larger sizes being obtained at lower product rates. However, the sensitivity of barite CSD to changes in product rate varies over the experimental range, and this is reflected in

Table 2 Summary of MSMPR/CLA Experimental Conditions and Parameters Derived from Barite Product CSD

Run No.	Product Rate (g/min)	Impellor Speed (RPM)	C_o^{++} (M Ba)	Vol. Avg Size, τ_a (μm)	Primary Growth Rate, G (μm/min)	Nucleation Rate, B^o (no./ml min)
1122	.272	200	.059	16.8 (.2)[b]	.075	6.0×10^2
1123*	.268	200	.055	18.5 (1.0)	.086	3.8×10^2
1221	.424	200	.086	15.7 (.2)	.080	9.1×10^2
1311+	.526	140	.113	16.4 (.2)	.095	1.3×10^3
1321	.508	200	.112	10.5 (.2)	.043	5.9×10^3
1322*	.504	200	.109	11.5 (.5)	.059	4.8×10^3
1331	.510	350	.109	9.5 (.2)	.046	9.4×10^3
1421	.796	200	.159	6.9 (.4)	.026	1.6×10^4
1511	.983	140	.193	5.8 (.1)	.021	4.3×10^4
1521	.994	200	.196	6.0 (.2)	.025	3.4×10^4
1531+	1.03	350	.210	7.6 (.05)	.038	3.7×10^4

* - Replication   + - Suspension Aging Test Performed

the primary growth rates (Table 2) which do not decrease steadily as product rate increases. In contrast, the nucleation rates do steadily increase with higher product rates. Again, this apparent anomaly can be attributed to agglomeration effects.

The irregularity of the growth rate trend with product rate can be explained by examining the agglomeration efficiency, β. Figure 3 depicts the calculated agglomeration efficiency for particle birth as a function of "captured" particle size, L', for two parametric values of "capturing" size, "L". Curves for three product rate conditions are shown. As product rate increases, it appears that β_b decreases much less markedly for the smaller capturing particles than for the larger ones. Since nucleation rates increase with the product rate, greater numbers of small particles are also available. Thus the net result of increased product rate is enhanced agglomeration; probably because higher local supersaturations at the higher precipitation rates permit more rapid intergrowth of barite crystals. The agglomerative mechanism acts to increase particle size, so that the total crystal surface area is less than if agglomeration were not present. Subsequently, the primary growth rate may not always be reduced as the nucleation rate increases.

Agitation intensity, as set by impellor speed, also influences barite CSD but to a smaller degree than product rate. In Figure 4, product CSD are shown for three impellor speed levels at a nominal product rate of 0.5 g/min. At this product rate there is a tendency toward smaller size as impellor speed increases. However, inspection of the primary growth and nucleation rate effects at a higher product rate reveals the influence of agglomeration.

Referring to Table 2, it is shown that at 0.5 g/min there is a significant decrease in primary growth rate, and an increase in nucleation rate, with respect to impellor speed. At a product rate of 1.0 g/min, the nucleation rate is essentially constant, while average particle size and primary growth rate actually increase with impellor speed.

The calculated effect of three impellor speed conditions on agglomeration efficiency for particle birth, β_b, appears in Figure 5. Comparison of these curves with those in Figure 3 suggests that, within the experimental ranges examined, agglomeration efficiency is more sensitive to product rate than agitation intensity. For this reason, it is plausible that under conditions which are favored by higher product rates, specifically, smaller sized particles and higher local supersaturations, agglomeration would be the predominant growth mechanism. Therefore, agglomeration effects could mask, and possibly reverse, expected primary growth and particle size trends with respect to agitation intensity.

Table 3 Agglomeration Kernel Model Discrimination

Run No.	Optimal SSQ Values				
	Power Law	Expon. Decay	Model II	Model I	Model III
1122	–	–	3.1	4.3	7.1
1123	–	–	1.6	5.7	4.1
1221	–	–	1.7	2.7	1.7
1311	–	–	2.5	4.3	1.6
1321	2.0	9.2	1.3	2.1	1.4
1322	2.5	3.2	2.9	3.9	2.4
1331	1.7	2.6	1.7	1.9	1.4
1421	–	–	7.6	10.2	5.8
1511	–	–	4.0	2.1	1.0
1521	1.9	11.1	2.0	2.1	3.1
1531	12.7	8.6	7.2	7.0	6.7
$\sum$ SSQ	20.8	34.7	36.3	46.3	36.5
Avg. SSQ	4.14	6.94	3.30	4.21	3.32
σ	4.80	3.81	2.19	2.59	2.27
σ^2	18.40	11.60	4.37	6.07	4.70
Conf. inter. on Avg. SSQ difference with Model II (.8 level)	–.10 1.76	1.21 6.07	–	–.39 2.20	–1.18 1.22

In conclusion, it has been found that a size-dependent agglomeration efficiency model is an essential component of a population balance model describing agglomerative barite growth. Robust empirical model forms can be developed from physical rationale, although the model derived in this study does not fully account for factors such as chemical environment and agitation effects. A more complex hydrodynamic model would be required to establish the fundamental kinetic constants for the agglomeration process.

Population balances incorporating this efficiency model provided an interpretive tool in discerning the role of agglomeration in formation of barite having a bimodal CSD. Both precipitation rate and agitation intensity affect the agglomerative growth mechanism of barite. Agglomeration of this sparingly soluble salt is apparently enhanced by more rapid cementing of contacted crystals at higher product rates. Since agglomeration efficiency is less sensitive to impellor speed than product rate, particle kinematics are apparently less important than the rate of crystal intergrown in successful formation of agglomerates.

NOTATION

B	= differential agglomerative particle birth rate (no./ml μm min)
B^o	= total particle nucleation rate (no./ml min)
CLA	= clear-liquor-advance
CSD	= Crystal Size Distribution
C_o	= cation feed concentration (M)
D	= differential agglomerative particle death rate (no./ml μm min)
G	= primary growth rate (μm/min)
k	= agglomeration model constant
L	= characteristic crystal dimension (μm)
L'	= captured particle size in agglomeration (μm)
L''	= capturing particle size in agglomeration, $L' < L''$, (μm)
L_o	= particle size at population balance initial condition (μm)
MSMPR	= Mixed-Suspension, Mixed-Product-Removal
n	= population density distribution function (no./ml μm)
n^o	= nuclei density parameter, population balance initial condition (no./ml μm)
P	= 80XY size analyzer channel number corresponding to largest particle size
Q_p	= volumetric product flow rate (ml/min)
R	= fines destruction parameter
SSQ	= sum-of-squares objective function
V	= crystallizer volume (ml)
β	= agglomeration efficiency function
γ	= agglomeration efficiency function model constant
λ	= agglomeration efficiency function model constant
σ	= variance
σ^2	= standard deviation
τ	= crystallizer solids residence time (min)

Subscripts

b	= particle birth
calc	= calculated value for n(L)
d	= particle death
data	= data value for n(L)
i	= 80XY size analyzer channel index

LITERATURE CITED

1. Maruscak, A. et al., Can. J. Chem. Eng., 49, 819-824 (1971).

2. Baker, C.G.J. and M.A. Bergougnou, Can. J. Chem Engr., 52, 246-250 (1974).

3. Halfon, A. and S. Kaliaguine, Can. J. Chem. Eng., 54, 168-172 (1976b).

4. Sarig, S. et al., J. Crystal Growth, 47, 365-372 (1979).

5. Hulbert, H.M. and S. Katz, Chem. Eng. Sci., 19, 555-574 (1964).

6. Nielsen, A.E., Acta Chem. Scan., 12, 951-958 (1958).

7. Nielsen, A.E., Acta Chem. Scan., 13, 1680-1686 (1959).

8. Gunn, D.J. and M.S. Murthy, Chem. Eng. Sci., 27, 1293-1313 (1972).

9. Nancollas, G.H. and N. Purdie, Trans. Faraday Soc., 59, 735-740 (1963).

10. Rizkalla, E.N., J. Chem. Soc. Faraday Trans. I, 79, 1857-1867 (1983).

11. Nancollas, G.H. and S.T. Liu, Soc. Petrol. Eng. J., 509-516 (December 1975).

12. Liu, S.T. et al., J. Crystal Growth, 33, 11-20 (1976).

13. Fischer, R.B., Anal. Chem., 23, 1667-1671 (1951).

14. Fischer, R.B. and T.B. Rhinehammer, Anal. Chem., 25, 1544-1548 (1953).

15. Kubota, N. and J.W. Mullin, J. Crystal Growth, 66, 676-678 (1984).

16. Halfon, A. and S. Kaliaguine, Can. J. Chem. Eng., 54, 160-167 (1976a).

17. Schumann, T.E.W., Quart. J. Roy. Meteor. Soc., 66, 195-207 (1940).

18. Thompson, P.D., "A Transformation of the Stochastic Equation for Droplet Coalescence," Proc. Int. Conf. on Cloud Physics, Toronto (August 1968).

19. Drake, R.L. "A general mathematical survey of the coagulation equation," in *Topics in Current Aerosol Research*, 3, Hidy, G.M. and J.R. Brock (Eds.) Pergamon Press, Oxford (1972).

20. Bapat, P.M. and L.L. Tavlarides, *AIChE J.*, 31, 659-666 (1985).

21. Ramabhadran, T.E. et al., *AIChE J.*, 22, 840-851 (1976).

22. Hooke, R. and T.A. Jeeves, *J. Assoc. Comp. Mach.*, 8, 212-229 (1961).

23. Kuester, J.L. and J.H. Mize, *Optimization Techniques with Fortran*, McGraw-Hill, New York (1973).

24. Randolph, A.D. and M.A. Larson, *Theory of Particulate Processes*, Academic Press, New York (1971).

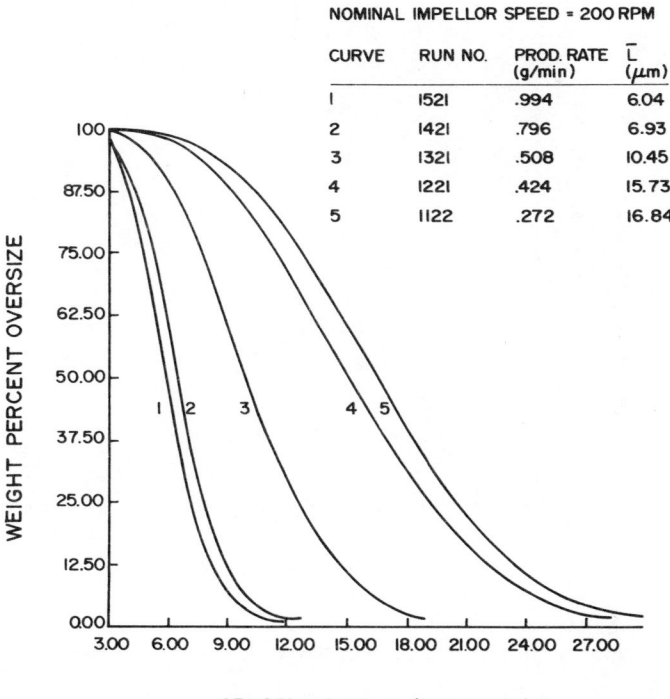

Figure 2. MSMPR/CLA product CSD—effect of barite product rate.

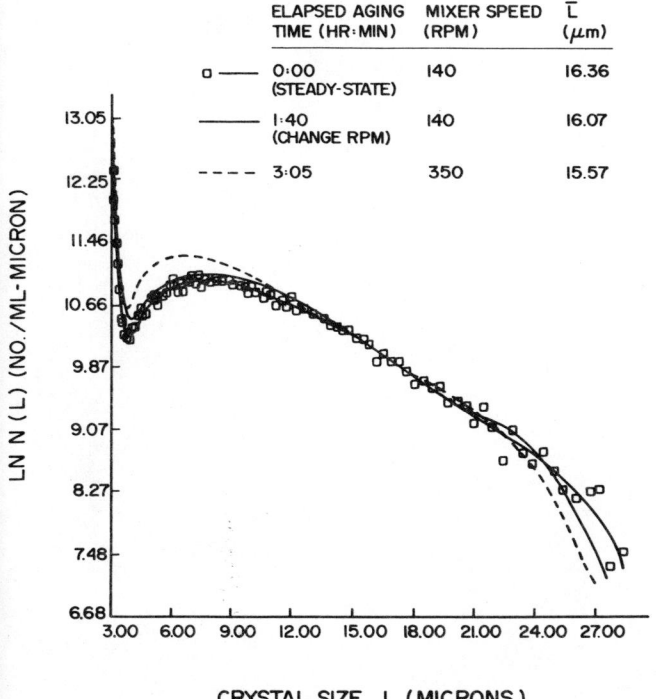

Figure 1. Results of suspension aging test for Run 1311—effect of increased agitation intensity.

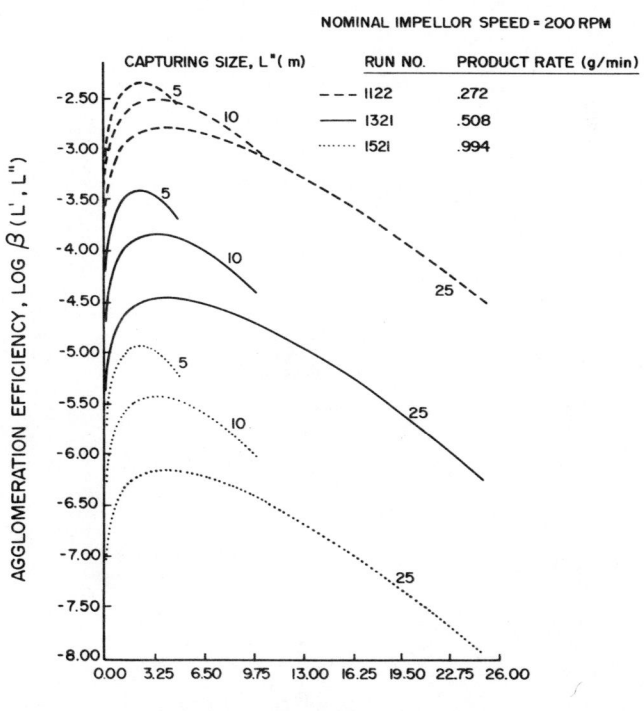

Figure 3. Calculated agglomeration efficiency for particle birth, β_b—effect of barite product rate.

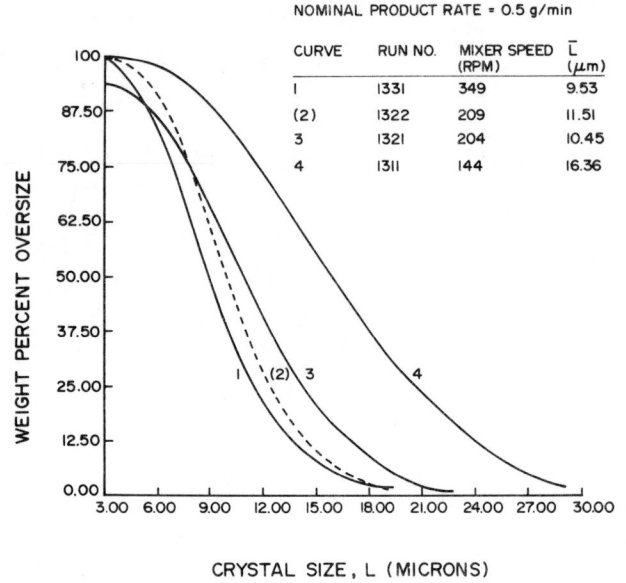

Figure 4. MSMPR/CLA product CSD—effect of impellor speed.

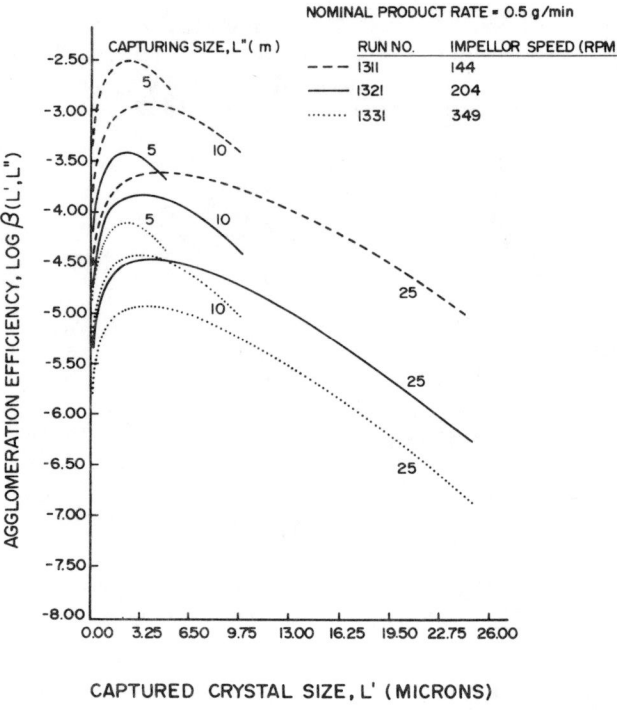

Figure 5. Calculated agglomeration efficiency for particle birth, β_b—effect of impellor speed.

CRYSTALLITE AGING UNDER TRANSPORT LIMITED CONDITIONS—APPLICATIONS OF MULTIPARTICLE DIFFUSION ALGORITHMS

S. P. Marsh and M. E. Glicksman ■ Rensselaer Polytechnic Institute, Troy, NY 12180-3590

The multi-particle diffusion problem (MDP) describes the diffusional interactions among an arbitrarily large number of domains or particles embedded in a continuous matrix phase. The quasi-static multiparticle diffusion field is modeled as an equivalent potential field. The particles are represented by point sources/sinks of heat or solute. A spatially uniform particle distribution is simulated by forming a basis of randomly located particles and repeating this basis in an infinite lattice framework. The resulting diffusion-field solution is obtained in both two and three dimensions by using an adaptation of Ewald's method for calculating lattice potential sums in a rapidly convergent form. The kinetic behavior of the system is numerically simulated via a time-stepping technique, using particle growth rates which include specific interparticle interactions. The simulation results for three-dimensional diffusion show deviations from the mean-field interaction analysis of Lifshitz, Slyozov and Wagner at moderate to high volume fractions. Two-dimensional simulation diffusion results indicate that significant departures from the predictions of Chakraverty's mean-field theory occur both at low and high surface coverages.

The properties and behavior of many inhomogeneous materials are affected by the simultaneous diffusional growth and dissolution of a dispersed second-phase embedded in a continuous matrix. Three-dimensional multiparticle diffusion, or "Ostwald ripening", occurs in such phenomena as the coarsening of fine solid structures in a solid-liquid mixture, the aging of precipitates from supersaturated solid solutions, and void growth during irradiation. Diffusion on a two-dimensional surface can occur in such practical processes as the sintering of supported metal catalysts and the aging of discontinuous thin films.

The driving forces for these reactions vary, but they generally involve gradients of temperature or concentration caused by the curvature-dependent solubility of the dispersed second-phase particles. A solution to the diffusion equation for the entire two-phase mixture is needed to model the kinetics of such reactions accurately. Many previous analyses of multiparticle diffusion problems (MDP's) rely on unrealistic assumptions to obtain approximations to the diffusion field. In this work we present a theoretical approach to both 2- and 3-dimensional MDP's which avoids many of these limiting assumptions by modeling the diffusion field as a dimensionless potential field. Lattice potential summation techniques, adapted from the method first used by Ewald [1], are used to obtain a detailed solution to the diffusion field which includes specific interparticle effects.

BACKGROUND

A major advance in the analysis of the 3-dimensional MDP is the continuum approach developed by Lifshitz and Slyozov [2] and Wagner [3]. In the LSW method, each individual spherical particle is assumed to react with a "mean field" located an infinite distance away. The diffusion field around the particle is found analytically, and the flux of heat or solute entering or leaving the particle is determined as a function of the particle size and "mean-field" value only. The "mean field" value is set by the average particle size of the ensemble of particles which is assumed to form a continuous size distribution. The hydrodynamic continuity equation is then applied to the particle size distribution (PSD) using the derived single-particle flux functions.

The relevant conclusions of this analysis are as follows:
1. The PSD becomes an asymptotic, invariant distribution when the particle radii are scaled by the average radius of the distribution.
2. The maximum radius of a particle in the asymptotic PSD is 1.5 times larger than the average radius.
3. The cube of the average radius of the PSD grows linearly with time at a scaled rate of 4/9.

A major drawback to this continuum approach is that it assumes a vanishingly small volume fraction of the dispersed second phase. This assumption allows individual particle fluxes to be calculated analytically but is physically unrealistic in most observed systems. A more detailed description of this analysis is described elsewhere [4], as are some attempts to modify the LSW approach to account for the volume-fraction effects, none of which are completely satisfactory.

Another approach to the MDP is a discrete summation technique, where the individual particle diffusion fields are summed to obtain the overall field. The earliest treatment of this method is that of Weins and Cahn [5]. Although it does account for specific particle interactions, the approach was limited to treating just a few particles. Therefore, their results cannot be extended reliably to provide statistical information on the coarsening kinetics of a large uniform dispersion of second-phase particles.

The continuum approach to the 2-dimensional MDP was developed by Chakraverty [6], in a manner analogous to the LSW method. The particles were modeled as hemispheres situated on a flat substrate. In this analysis, however, the radially symmetric diffusion field for each individual particle extends to a "screening distance" equal to e (2.718...) times the particle radius, where the concentration is assumed to reach its uniform average value. This "screening distance" is infinite in the LSW approach, but must be made finite here because the single-particle potential in two dimensions is logarithmic and therefore becomes divergent at large distances. Using the finite screening distance assumption, Chakraverty determined the flux entering or leaving a particle in terms of the particle size. Applying the hydrodynamic continuity equation to the 2-dimensional PSD using this flux function, he concluded the following:

1. The PSD approaches a time-independent distribution when the particle radii are scaled by the average radius of the ensemble (as in the LSW case).
2. The maximum radius of a particle in this asymptotic distribution is 4/3 as large as the average radius.
3. The fourth power of the average radius grows linearly with time, at a scaled rate of 27/64.

Chakraverty's analysis is outlined in more detail elsewhere [7]. The main drawback here is the artificial "screening distance" imposed on each particle, since the calculated particle fluxes are sensitive to this distance. Again, specific particle interaction effects were ignored as were the effects of fractional surface coverage on diffusion rates.

ANALYSIS

A detailed description and solution of the multiparticle diffusion field will now be developed. Physically, a collection of spherical particles in a three-dimensional matrix or a collection of hemispheres on a flat substrate will continuously absorb and emit heat or solute atoms at their boundaries. In such a system where the matrix served only to conduct heat or mass fluxes, the larger particles tend to grow while the smaller ones tend to shrink. This behavior results from the application of the Gibbs-Thomson or Thomson-Freundlich boundary condition to each particle. For the case of mass transfer, the Thomson-Freundlich condition states that smaller particles are in equilibrium with a relatively higher solute atom concentration at their boundaries because of their high curvature. Mass tends to diffuse away from these smaller particles, causing them to grow even smaller. Solute atoms tend to diffuse toward larger particles because their equilibrium perimeter concentration is relatively lower. Similar arguments hold for the case of heat transfer, with temperature gradients in the matrix leading to simultaneous growth and dissolution of particles.

From this discussion it can be seen that each particle affects the growth rate of nearby particles by influencing the availability of diffusing heat or solute. Each particle is thus a source or sink of heat or solute, depending on whether it is growing or shrinking at a given moment. In potential terms, each particle can be modeled as an equivalent point source of strength B, where B is proportional to the volumetric growth rate of a particle. B will be negative for a shrinking particle, and will be a function of time for any given particle as it and the surrounding particles change in size.

If the ripening process is diffusion-controlled, so that interfacial absorption/desorption is not the rate-limiting step, then the quasi-static diffusion field can be described by Poisson's equation. The variables in the diffusion equation can be appropriately scaled for either heat or mass diffusion [7,8] so that θ represents a dimensionless concentration or temperature. The resulting form of Poisson's equation which describes the diffusion field arising from a collection of N point sources/sinks is as follows:

$$\nabla^2 \theta = \sum_{j=1}^{N} -2\pi B_j \delta(\vec{r} - \vec{r}_j) \quad [\text{2-D}], \quad (1a)$$

and
$$\nabla^2 \theta = \sum_{j=1}^{N} -4\pi B_j \delta(\vec{r} - \vec{r}_j) \quad [\text{3-D}]. \quad (1b)$$

In these equations B_j is the source strength of the jth particle, r_j is the location of the center of the jth particle, $\vec{r}$ is an arbitrary field point, and δ is the Dirac delta function.

The source strength B_j represents the dimensionless volumetric growth rate of the jth particle. For a system of N particles which exchange mass or heat through the matrix, where there are no external sources or sinks, the mass/heat conservation equation can be written as

$$\sum_{j=1}^{N} B_j = 0. \quad (2)$$

This equation is equivalent to the statement that the total volume of second-phase particles remains constant with time.

Based on the linearity of Poisson's equation, a solution to equations (1a) and (1b) can be written as the linear combinations

$$\theta(\vec{r}) = B_o + \sum_{j=1}^{N} B_j \log |\vec{r} - \vec{r}_j| \quad [\text{2-D}], \quad (3a)$$

and
$$\theta(\vec{r}) = B_o + \sum_{j=1}^{N} \frac{B_j}{|\vec{r} - \vec{r}_j|} \quad [\text{3-D}], \quad (3b)$$

where B_o is an arbitrary, constant reference potential. The multiparticle diffusion field can be calculated straightforwardly from equations (3a) and (3b), but two difficulties arise when applying this method to an arbitrarily large number of particles. The first is that the summation in equation (3a) is divergent for large N because of the logarithmic terms, whereas the summation in equation (3b) is semi-convergent, and the limit can depend on the order in which the terms are summed. A second problem with attempting direct summation is that a scheme is needed to specify the locations of an arbitrarily large number of uniformly dispersed particles.

These difficulties can be averted by creating a finite "basis" and a repeating lattice vector, $\vec{r}_e$. This basis will be a square for 2-D diffusion, and a cube for the 3-D case. The basis is filled with a finite number of randomly located particles arranged in the desired size distribution. The size of the basis is then determined by the specified surface coverage (2-D) or volume fraction (3-D). The basis is then repeated infinitely in two or three dimensions to fill space in a Bravais lattice configuration. Each lattice translation vector, $\vec{r}_e$, locates the origin of a translated image of the basis in the Bravais lattice. A two-dimensional diagram of this configuration is pictured in Figure 1, where the basis and eight surrounding translations are shown. In this figure, each cell contains 29 particles.

The solution to Laplace's equation in two or three dimensions, given by equations (3a) and (3b), can be reformulated in terms of the basis and lattice translation vector as

$$\theta(\vec{r}) = B_o + \sum_{\vec{r}_e} \sum_{i=1}^{N'} B_i \log |\vec{r} - \vec{r}_i - \vec{r}_e| \quad [\text{2-D}] \quad (4a)$$

$$\theta(\vec{r}) = B_o + \sum_{\vec{r}_e} \sum_{i=1}^{N'} \frac{B_i}{|\vec{r} - \vec{r}_i - \vec{r}_e|} \quad [\text{3-D}]. \quad (4b)$$

where N' is the number of particles in the basis. The advantage to this reformulation is two-fold. First, a uniform spatial distribution containing an infinite number of particles is completely specified by the locations of only N' particles and the size of the basis. Secondly, equations (4a) and (4b) now resemble lattice potential equations, for which there exist mathematical transformations which convert the lattice sums into rapidly and absolutely convergent summations. This technique was first developed by Ewald [1] to determine the coulombic potential in alkali halide structures. Although the source strengths in an ionic lattice are known and constant, whereas the B_i's in a diffusion problem are as yet undetermined and vary with time, the Ewald technique can still be applied to equations (4a) and (4b) to obtain a rapidly convergent form of the dimensionless potential field $\theta(\vec{r})$. The adaptation of Ewald's method to this problem is described in detail elsewhere [4,7], and results in the following equations:

$$\theta(\vec{r}) = B_0 + \left(\frac{2\pi}{a_o^2}\right) \sum_{i=1}^{N'} \sum_{\vec{g}}{}' \frac{B_i}{G^2} \exp\left[\frac{-G^2 a_o^2}{4\pi} + i \vec{g} \cdot (\vec{r} - \vec{r}_i)\right]$$

$$- \frac{1}{2} \sum_{i=1}^{N'} \sum_{\vec{r}_e} B_i \, Ei\left[-\frac{\pi}{a_o^2} |\vec{r} - \vec{r}_i - \vec{r}_e|^2\right] \quad \{2\text{-D}\} \quad , \quad (5a)$$

$$\theta(\vec{r}) = B_0 + \left(\frac{4\pi}{a_o^3}\right) \sum_{i=1}^{N'} \sum_{\vec{r}_e} \frac{B_i}{G^2} \exp\left[\frac{-G^2 a_o^2}{4\pi} + i \vec{g} \cdot (\vec{r} - \vec{r}_i)\right]$$

$$+ \sum_{i=1}^{N'} \sum_{\vec{r}_e} \frac{B_i \, \text{erfc}\left[\left(\frac{\pi}{a_o^2}\right)^{1/2} |\vec{r} - \vec{r}_i - \vec{r}_e|\right]}{|\vec{r} - \vec{r}_i - \vec{r}_e|} \quad [3\text{-D}] \quad . \quad (5b)$$

These equations allow the dimensionless potential θ to be calculated from the $N'+1$ source strengths B_i (including B_0), and the locations of the N' particles in the basis. The B_i values, which are proportional to the growth rates of the particles, are as yet undetermined. Their values can be found by applying the Gibbs-Thomson boundary condition to each particle. The linearized Gibbs-Thomson equation can be written in terms of dimensionless variables as

$$\theta(R_i) = \frac{1}{R_i} \quad \text{(mass transfer)}, \quad (6a)$$

or

$$\theta(R_i) = -\frac{1}{R_i} \quad \text{(heat transfer)}, \quad (6b)$$

for both the 2- and 3-dimensional cases considered, where R_i is the dimensionless radius of the ith particle.

This condition can be strictly applied to the case of a single-particle diffusion field, which is radially or spherically symmetric around the particle. In the more complex multiparticle diffusion field, the potential θ will vary over the perimeter of a hemispherical particle (2-D) or the surface of a sphere (3-D). This boundary condition can be applied, however, if it is equal to the perimeter- or surface-averaged potential, $\bar{\theta}$. In both two and three dimensions, this is equivalent to setting the potential at the center of the ith particle, $\theta(\vec{r}_i)$, equal to $1/R_i$. Applying this averaged boundary condition to the ith particle and using equations (5a) and (5b) for $\theta(\vec{r})$ leads to the following relation between B_i and R:

$$\frac{1}{R_i} = B_0 + B_i \log R_i + \left(\frac{2\pi}{a_o^2}\right) \sum_{j=1}^{N'} \sum_{\vec{g}}{}' \frac{B_j}{G^2} \exp\left[\frac{-G^2 a_o^2}{4\pi} + i \vec{g} \cdot (\vec{r}_j - \vec{r}_i)\right]$$

$$+ \left(\frac{1}{2}\right) \sum_{\substack{j=1 \\ (j \neq i)}}^{N'} \sum_{\vec{r}_e} B_j \, Ei\left[-\frac{\pi}{a_o^2} |\vec{r}_i - \vec{r}_j - \vec{r}_e|^2\right]$$

$$+ \left(\frac{B_i}{2}\right) \left(\gamma + \log\left[\frac{\pi}{a_o^2}\right] + \sum_{\vec{r}_e}{}' Ei\left[-\frac{\pi}{a_o^2} |\vec{r}_e|^2\right]\right) \quad [2\text{-D}] \quad (7a)$$

$$\frac{1}{R_i} = B_0 + B_i \left[\frac{1}{R_i} - \frac{2}{a_o}\right] + \left(\frac{4\pi}{a_o^3}\right) \sum_{j=1}^{N'} \sum_{\vec{g}}{}' \frac{B_j}{G^2} \exp\left[\frac{-G^2 a_o^2}{4\pi} + i \vec{g} \cdot (\vec{r}_j - \vec{r}_i)\right]$$

$$+ \sum_{j=1}^{N'} \sum_{\vec{r}_e} \frac{B_i \, \text{erfc}[|\vec{r}_j - \vec{r}_i - \vec{r}_e| \, (\frac{\pi}{a_o^2})^{1/2}]}{|\vec{r}_j - \vec{r}_i - \vec{r}_e|} \quad [3\text{-D}] .$$
$$\begin{pmatrix}\vec{r}_e = 0 \\ i \neq j\end{pmatrix} \quad (7b)$$

In Equation (7a), γ represents Euler's constant.

There exist N' independent equations of the type shown in (7a) and (7b), which are obtained by applying the Gibbs-Thomson boundary condition averaged over each particle in the basis. However, one more equation is needed to solve for the $N'+1$ unknowns $B_0 - B_{N'}$. This last equation is a modification of the mass/heat conservation condition of equation (2). Because all translated images of the basis behave identically, the conservation of mass or heat can be restated as a local requirement, i.e.,

$$\sum_{i=1}^{N'} B_i = 0, \quad (8)$$

where the summation is carried out over just the particles in the basis.

There are now $N'+1$ unknowns, and $N'+1$ independent equations that relate them. These equations are the N' boundary conditions (7a, 7b) and the local conservation condition (8). To determine the source strengths B_i, these equations can be manipulated to form the linear system.

$$\{Y_i\} = [X]\{B_i\}, \quad (9)$$

where

$$\{Y_i\} = \begin{bmatrix} \pm \frac{1}{R_1} \\ \pm \frac{1}{R_2} \\ \cdot \\ \cdot \\ \cdot \\ \pm \frac{1}{R_{N'}} \\ 0 \end{bmatrix} \quad \text{and} \quad \{B\} = \begin{bmatrix} B_1 \\ B_2 \\ \cdot \\ \cdot \\ \cdot \\ B_{N'} \\ B_0 \end{bmatrix} \quad (10)$$

The symmetric coefficient matrix [X] can be written as follows:

$$[X] = \begin{bmatrix} D_{11} & A_{12} & A_{13} & \cdots & 1 \\ A_{21} & D_{22} & A_{23} & \cdots & 1 \\ A_{31} & A_{32} & D_{33} & \cdots & 1 \\ \cdot & \cdot & \cdot & \cdots & \cdot \\ \cdot & \cdot & \cdot & \cdots & \cdot \\ \cdot & \cdot & \cdot & \cdots & \cdot \\ 1 & 1 & 1 & \cdots & 0 \end{bmatrix} \quad (11)$$

The terms appearing on the main diagonal are defined as:

$$D_{ii} = \log R_i + \left(\frac{2\pi}{a_o^2}\right) {\sum_{\vec{g}}}' \frac{1}{G^2} \exp\left[\frac{-G^2 a_o^2}{4\pi}\right]$$
$$+ \left(\frac{1}{2}\right) {\sum_{\vec{r}_e}}' \text{Ei}\left[-\frac{\pi}{a_o^2}|\vec{r}_e|^2\right] + \frac{1}{2}\left(\gamma + \log\frac{\pi}{a_o^2}\right) \quad [\text{2-D}], \quad (12a)$$

$$D_{ii} = \frac{-2}{a_o} + \frac{4\pi}{a_o^2}\left[{\sum_{\vec{g}}}' \frac{1}{G^2} \exp\frac{-G^2 a_o^2}{4\pi}\right] + {\sum_{\vec{r}_e}}' \frac{\text{erfc}\left[|\vec{r}_e|\left(\frac{\pi}{a_o^2}\right)^{\frac{1}{2}}\right]}{|\vec{r}_e|}$$
$$[\text{3-D}], \quad (12b)$$

and the off-diagonal terms A_{ij} are defined as

$$A_{ij} = A_{ji} = \left(\frac{2\pi}{a_o^2}\right) {\sum_{\vec{g}}}' \frac{1}{G^2} \exp\left[\frac{-G^2 a_o^2}{4\pi} + i\vec{g}\cdot(\vec{r}_j - \vec{r}_i)\right]$$
$$+ \left(\frac{1}{2}\right) \sum_{\vec{r}_e} \text{Ei}\left[-\frac{\pi}{a_o^2}|\vec{r}_j - \vec{r}_i - \vec{r}_e|^2\right] \quad [\text{2D}] \quad (13a)$$

$$A_{ij} = A_{ji} = \left(\frac{4\pi}{a_o^3}\right)\left[{\sum_{\vec{g}}}' \frac{1}{G^2} \exp\frac{-G^2 a_o^2}{4\pi} + i\vec{g}\cdot(\vec{r}_j - \vec{r}_i)\right]$$
$$+ \sum_{\vec{r}_e} \frac{\text{erfc}\left[|\vec{r}_j - \vec{r}_i - \vec{r}_e|\left(\frac{\pi}{a_o^2}\right)^{\frac{1}{2}}\right]}{|\vec{r}_j - \vec{r}_i - \vec{r}_e|}$$
$$[\text{3-D}] \quad (13b)$$

These matrix elements, D_{ii} and A_{ij}, contain rapidly convergent sums which result from the application of the Ewald transformation to the multiparticle diffusion problem. The coefficient matrix [X] may be termed the <u>particle interaction matrix</u>, since it relates the source strength B of each particle in the basis to the source strength of every other particle in the basis <u>and all particle images in the translated cells</u> through the averaged Gibbs-Thomson boundary conditions. The net result of this analysis is that equation (9), with the definitions given in equations (10)-(13), provides a unique determination of the diffusion-controlled growth rate of each particle in the model, based on the radii and locations of these particles in the basis. These growth rates are calculated directly from the 2- and 3-dimensional multiparticle diffusion fields, which are found without a recourse to the limiting assumptions of either low volume fraction, a finite number of particles, or of a fixed "screening distance".

In order to obtain information on the kinetics of diffusion-controlled multiparticle ripening, the source strengths B must be explicitly related to the particle growth rates. Based on the non-dimensionalizing parameters and the definition of B in equation (1), it can be shown that for both the 2-D and 3-D cases,

$$\dot{R}_j = \frac{B_j}{R_j^2}, \quad (14)$$

where $\dot{R}_j \equiv dR_j/dt$. The ripening behavior is then simulated by numerically solving equation (9) for the instantaneous B values and choosing a small time interval Δt for evolution of the system. The changing particle sizes were found by using a linearization of equation (14) of the form

$$R_j(t+\Delta t) = \frac{B_j(t)}{R_j^2(t)} \Delta t. \quad (15)$$

This time-stepping procedure can be continued by alternately using equation (9) to find the B values at the beginning of each time interval and then applying equation (15) to calculate the new particle radii at the end of each time interval Δt. The particle size distribution in the basis will thus evolve with time, with the larger particles tending to grow while the smaller ones shrink. When a particle shrinks below a certain minimum radius, it is assumed to have completely dissolved and is removed from the particle distribution. This is accomplished mathematically by eliminating the row and column corresponding to the dissolved particle from the interaction matrix [X], thereby reducing the number of simultaneous equations by one. All other elements in the matrix remain unchanged because the locations of the other particles in the basis remain fixed by assumption.

The use of Laplace's equation (1) to describe the multiparticle diffusion field requires that the field be quasi-static. This assumption can be justified by choosing the time interval Δt small enough so that the changes in the particle radii over the time-step are relatively small. Each time-step should also be large enough to allow significant ripening to occur in the simulation over a reasonable number of time-steps.

The number of particles initially placed in the basis, N', should be as large as possible to provide maximum spatial randomness and a statistically smooth size distribution in the model. This number is limited by the computational time required to solve the (N'+1) simultaneous equations for the B values at each time-step.

Using the above guidelines, computer simulations of multiparticle ripening have been carried out in both two and three dimensions for a variety of conditions. For the two-dimensional case, simulation results indicate the particle source strengths vary significantly from those predicted by Chakraverty's mean-field analysis. Attempts have been made to vary the screening distance with surface coverage to provide a better analytical fit to the simulation data [7]. An example of this is shown in Figure 2 where particle source strengths are plotted against particle size at a surface coverage of 1.0%.

More extensive computer simulations have been carried out using the three-dimensional diffusion-field model. The results have been detailed elsewhere [9], but the main conclusions may be summarized as follows: Steady-state particle-size distributions have been obtained for a range of volume fractions. These asymptotic distributions, obtained by statistically averaging a number of different simulations, are shown in Figure 3. The relatively narrow LSW distribution was obtained from the limiting case of zero volume fraction (Figure 4). As the volume fraction of the dispersed phase increases, the distribution becomes broader and somewhat flatter. Over the range of volume fractions examined, the cube of the average radius of the distribution was found to grow linearly with time once a steady-state distribution was attained. The scaled rate constant for this process approached the LSW value of 4/9 at zero volume fraction as shown in Figure 5, and increased significantly with increasing volume fraction. These rate constants have been compared with those obtained by other continuum ripening theories [9], but unfortunately there is insufficient experimental data at present to differentiate between these theoretical results.

SUMMARY AND CONCLUSIONS

The aging of a dispersed two-phase material often occurs by the diffusion of heat or mass among the second-phase particles. The present work describes models for this diffusion-limited ripening which determine the growth rate of each particle in a uniform dispersion directly from the diffusional interactions among the particles. The following is a summary of the multiparticle diffusion model presented in this paper:

1. A finite basis cell is formed which contains N' randomly dispersed particles. In 2-dimensional diffusion, this basis cell is a square plane with hemispherical particles on the surface. The three-dimensional basis is a cube with spherical particles scattered throughout it. The basis cell is then repeated in two or three dimensions as a regular Bravais lattice to form an infinite, uniform dispersion of particles.

2. Each ripening particle is modeled as an equivalent point source/sink of mass or heat. The resulting diffusion field is equivalent to a dimensionless potential field. Ewald's method is used to derive mathematically convergent forms of both the two- and three-dimensional multiparticle fields.

3. The perimeter- or surface-averaged Gibbs-Thomson boundary condition is applied to each of the N' particles in the basis. This results in N' equations relating the source strengths to the particle sizes and locations in the basis. The overall conservation of mass or heat is used to create one more equation relating the source strengths.

4. The (N'+1) equations form a linear system which can be solved to determine all of the particle source strengths at any point in time.

5. The growth rate of each particle is mathematically related to its source strength by considering the flux of mass or heat at the particle perimeter or surface. The resulting growth rates are then used in a time-stepping scheme to simulate the collective ripening of all particles in the system.

The following conclusions can be made based on the ripening model presented in this paper:

1. The multiparticle diffusion field can be recast into a mathematically convergent form through the use of lattice potential techniques.

2. Particle growth rates determined by this model arise directly from the diffusional interactions between all particles in a uniform dispersion. Volume-fraction effects are implicitly accounted for through the size and configuration of the basis used in the model.

3. Two-dimensional diffusion simulations indicate that the asymptotic particle-size distribution will generally be broader than that predicted by Chakraverty, who used a fixed "screening distance" assumption to determine the individual particle growth rates.

4. Extensive computer simulations in three dimensions verify that the LSW distribution and rate constant are obtained in the limiting case of zero volume fraction of the dispersed second phase. Volume-fraction effects are found which indicate that broader asymptotic size distributions and higher rate constants result at higher volume fractions.

ACKNOWLEDGEMENTS

The authors gratefully acknowledge support of this work by the National Science Foundation under contract DMR86-11302.

REFERENCES

1. P.P. Ewald, Ann. Physik, (4) 64, 253 (1921).

2. I.M. Lifshitz and V.V. Slyozov, J. Phys. Chem. Solids, 19, 35 (1961).

3. C. Wagner, Z. Electrochem., 65, 581 (1961).

4. P.W. Voorhees, Ph.D. Thesis, Rensselaer Polytechnic Institute (1982).

5. J.J. Weins and J. Cahn, in Sintering and Related Phenomena, ed. by G.C. Kuczynski, p. 151, Plenum, NY (1973).

6. B.K. Chakraverty, J. Phys. Chem. Solids, 28, 2401 (1967).

7. S.P. Marsh, M.S. Thesis, Rensselaer Polytechnic Institute (1984).

8. P.W. Voorhees and M.E. Glicksman, Acta. Metall., 32, 2001 (1984).

9. P.W. Voorhees and M.E. Glicksman, Acta. Metall., 32, 2013 (1984).

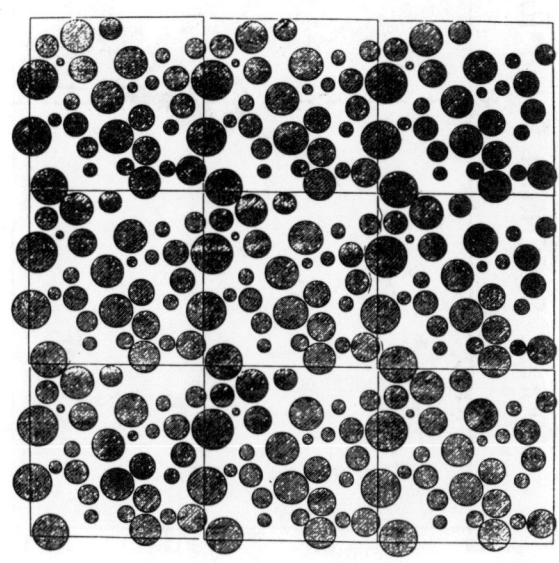

Figure 1 System of particles consisting of a basis and eight surrounding translated cells.

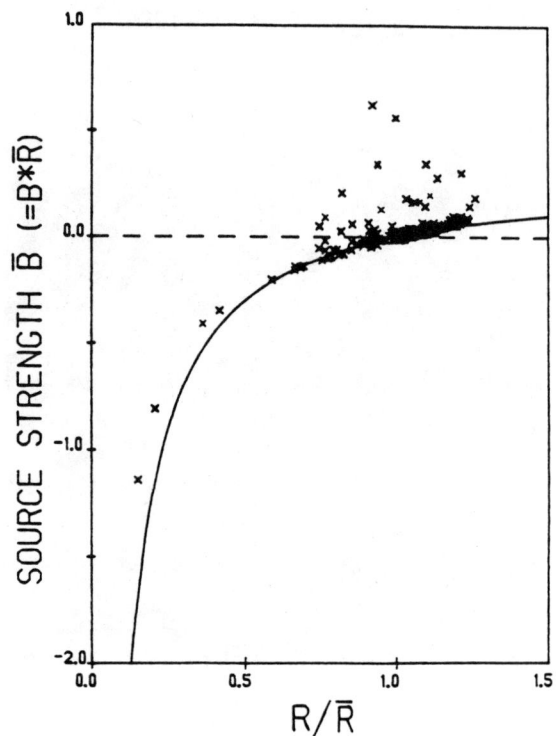

Figure 2 Particle source strengths obtained from two-dimensional simulations at 1.0% surface coverage. The dashed curved line indicates Chakraverty's mean-field source strength function, while the solid curve represents a modified mean-field function using a screening distance equal to 29 times the particle radius.

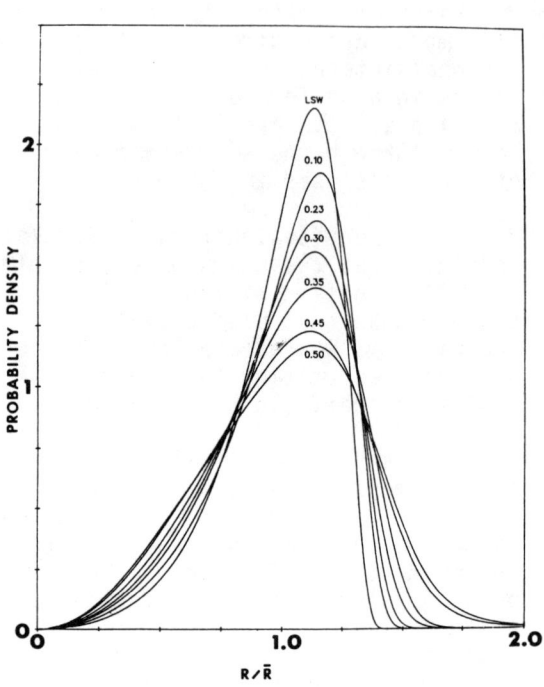

Figure 3 Steady-state distributions at various volume fractions for three-dimensional diffusion. These curves were obtained by statistically averaging and smoothing discrete histograms obtained in a number of separate simulations.

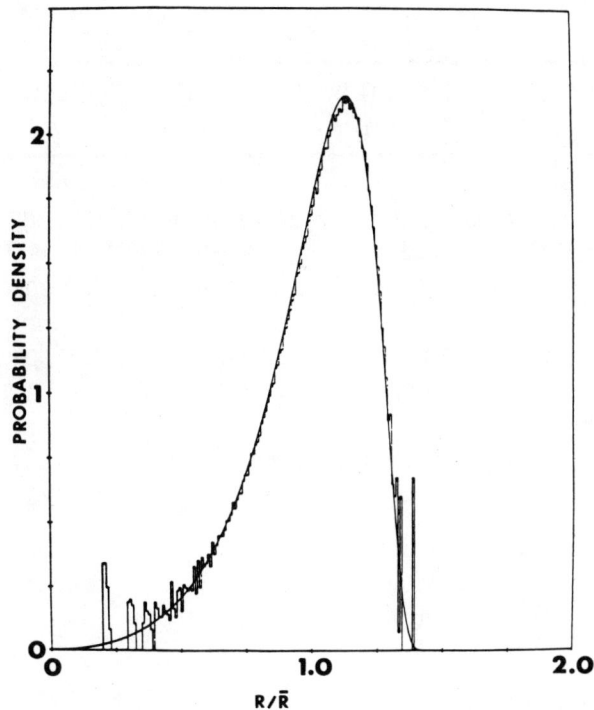

Figure 4 Steady-state analytical distribution from LSW theory and simulation derived histogram at $f_v=0$. Noise in histogram near the tails of the distribution occur because of the finite number of basis particles used in the simulation.

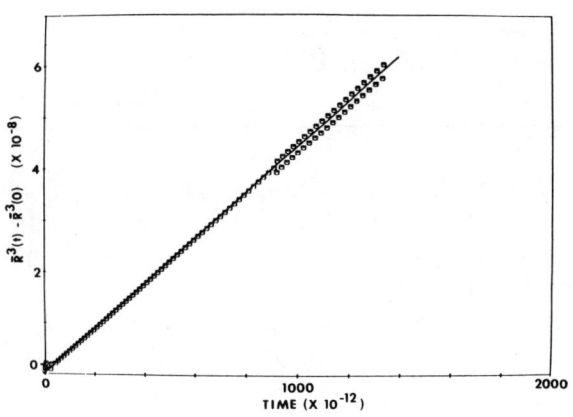

Figure 5 Cube of the average particle size vs. time at zero volume fraction. Straight line indicates the LSW slope of 4/9, which agrees closely with simulation data.

STATE SPACE REPRESENTATION OF THE DYNAMIC CRYSTALLIZER POPULATION BALANCE: APPLICATION TO CSD CONTROLLER DESIGN

Syuji Tsuruoka and A. D. Randolph ■ College of Engineering and Mines, Department of Chemical Engineering, University of Arizona, Tucson, AZ 85721

The dynamic population balance equation (PBE) representing crystal-size distribution (CSD) from mixed magma crystallizers with size-dependent removal rates is a nonlinear hyperbolic wave equation with size-dependent coefficients. This PBE representing crystallizers with arbitrarily complex configurations was cast in a vector-matrix state space formulation suitable for CSD dynamics.

Previously determined CSD stability limits for idealized MSMPR and R-z crystallizers were verified with this state space model. The state space formulation is model-independent, but the R-z crystallizer model was used to load the coefficient matrix to simulate CSD dynamics from an experimental KCl crystallizer.

Finally, an on-line CSD controller was illustrated using a minimum order Luenberger observer control algorithm. This control technique should be feasible using currently available light-scattering size measurement technology in a conditioned fines stream.

INTRODUCTION

The dynamic crystal size distribution (CSD) equations were cast in a state space formulation using the method of lines. This algorithm does not need any particular crystallizer configuration nor nucleation/growth rate kinetics. The dynamic population balance equation (which is a PDE) was converted to ODE's and thus conventional analytical methods in systems theory could be adopted to analyze the crystallizer to design a controller. Simulated stability results for classified (the so-called R-z configuration) and MSMPR crystallizers were identical to those developed by Randolph et al. (1973). The new algorithm should serve as a versatile model-independent simulator for crystallization systems.

DEVELOPMENT OF MODEL

The starting point of the mathematical development is the following set of equations (Randolph et al. (1973)) generally recognized to represent CSD dynamics in crystallizers of complex configuration, e.g. with fines dissolving and classified product removal (the R-z crystallizer).

$$\frac{\partial u}{\partial \theta} + \phi \frac{\partial u}{\partial \chi} + hu = 0 \qquad (1)$$

University of Arizona, Tucson, Arizona, S. Tsuruoka is now with Nippon Lever B.V., Tokyo, Japan.

$$\phi = \frac{1+\beta f_{3,1}}{1+\beta} \cdot \frac{1}{f_2} \qquad (2)$$

$$U_o = f_3^j \phi^{i-1} \qquad (3)$$

where the variables have been rendered dimensionless and normalized by the following transformations:

Time $\qquad \theta = \dfrac{t}{V/Q_p}$

Population density $\qquad u = \dfrac{nG^\circ}{B^\circ}$

Growth rate $\qquad \phi = \dfrac{G}{G^\circ}$

Particle size $\qquad \chi = \dfrac{LQ_p}{VG}$

Removal rate $\qquad h = \dfrac{Q(L)}{Q_p}$

Fraction mass dissolved

$$\beta = \frac{(R-1)}{6D_p} \int_o^{x_F} x^3 \bar{u}\, dx$$

where D_p is a dimensionless function of the third order gamma distribution evaluated at the fines and product classification sizes, (Randolph and Larson, p145).

Normalized fractional moment

$$f_{j,k} = \frac{\int_0^{L_k} L^j n \, dL}{\int_0^{L_k} L^{\bar{j}} n \, dL}$$

and where

$$h(L) \equiv Q(L)/Q = \begin{cases} R, & x < x_F \\ 1, & x_p \text{ in } (x_F, x_p) \\ z, & x > x_p \end{cases}$$

which represents the size-dependent particle removal rate in the R-z crystallizer. Equation (1) is converted into a semi-discrete approximate formula using a fourth-order correct central difference. Thus,

$$\frac{du_\ell}{d\theta} + \phi \frac{u_{\ell-2} - 8u_{\ell-1} + 8u_{\ell+1} - u_{\ell+2}}{12\delta} + h_\ell u_\ell = 0 \quad (4)$$

Two fictitious grid points are needed to approximate the boundary condition for Equation (4). The sequential scheme of the numerical calculation is as follows. Suppose that all grid points, including fictitious points u_{-1} and u_{-2}, are known at time = 2. These known points give the next time step points with a fourth order correct approximation using 4 points. Nuclei density u_0 is independently calculated (from Equation (3)) using all of the real points at time = 2 and replaced at time = 3. Then, fictitious points are calculated using the new grid points at time = 3 and so forth.

Using a vector-matrix form, Equation (1) becomes:

$$\dot{\underline{u}}(\theta) = A(\theta) \underline{u}(\theta) \quad (5)$$

where vector $\underline{u}$ is:

$$\underline{u} = (u_0, u_1, \ldots, u_N)' \quad (6)$$

and A is as shown in Figure 1. The form of matrix A is pentadiagonal as seen from Figure 1. Thus, matrix A is sparse, permitting rapid calculations, but can be of high order to cope with the parameter sensitivity of Equation (1).

Equation (5) is of homogeneous state variable form with a slowly changing time variant matrix and is useful as it stands for direct simulation of CSD dynamics and demonstration of system stability. To implement numerical calculations, Equation (5) can be arranged in a general form with an external input, represented as:

$$\dot{\underline{u}}(\theta) = A(\theta) \underline{u}(\theta) + B\underline{r}(\theta) \quad (7)$$

where B is the elementary matrix and $\underline{r}(\theta)$ represents external perturbations to the system.

An analytical solution of Equation (7) can be obtained by formal integration (Vemuri and Karplus (1981)). Nevertheless, according to Kailath (1980), it is impossible to solve the equation analytically unless the matrix A is unitary. Thus, the equation is converted to discrete form as:

$$\underline{u}_{m+1} = \Phi_{m+1} \underline{u}_m + \int_0^{\theta m} \Phi_m(\theta - a) B\underline{r}_m d\sigma \quad (8)$$

with

$$\Phi_m = \exp[A_m \Delta \theta] \quad (9)$$

If the time increment $\Delta\theta$ is small enough and unique, Equations (8) and (9) can be rewritten as:

$$\underline{u}_{m+1} = \Phi_{m+1} \underline{u}_m + \Delta\theta \frac{(\Phi_m B\underline{r}_{m+1} + \Phi_m B\underline{r}_m)}{2} \quad (10)$$

where

$$\Phi_m = \exp(A_m \Delta\theta) \quad (11)$$

The mathematical derivation is detailed elsewhere (Tsuruoka (1986)).

SIMULATED RESULTS

Simulations were implemented using the new algorithm as described below. Figure 2 shows simulation results for an MSMPR crystallizer with various nuclei sensitivity values i. The critical stability point for an MSMPR crystallizer is $i_c = 21$. It is evident that these simulation results show convergence for $i < 21$ and divergence for $i > 21$. Figure 3 shows the simulation results for an R-z crystallizer with various nuclei sensitivity values. In this case, the critical point is $i_c = 3.8$ (Randolph et al. (1973)). The dynamic CSD curves are consistent with this stability criterion. The new simulation algorithm thus satisfies these criteria and can be assumed to numerically predict dynamic CSD in a continuous crystallizer.

To show simulation for an experimental crystallizer, a set of data by Randolph et al. (1986) was used. A sketch of the apparatus is illustrated in Figure 4. The KCl crystallizer consists of double draw-off product removal system. There exists no product classification device, but the product was apparently classified internally, as evident from their experimental data. A fictitious product classifier was assumed. Experimental data are summarized in Table 1. The adjustable parameters necessary to fit these data, including the upset configuration, are summarized in Table 2.

Xp and z were chosen as 10 and 1, or 6 and 5 respectively. Choosing these numbers determines the slope of the semi-log distribution. The upset was assumed to be a rectangular pulse. Simulation results are shown in Figures 5(a) and (b) at two different crystal sizes. It is evident that in general the contours of the pulse propagation correspond to those of the experimental results except for their intensities. The reason for this discrepancy is that the initial steady state CSD in the experiment was not an ideal R-z distribution, yet idealized R-z parameters were used for the simulation. The dynamic CSD of an R-z crystallizer could be predicted using the new algorithm given a proper set of parameters: R, z, X_F, X_p, and the correct upset configuration. The transit time of the pulse between the two sizes shown was of course correctly modelled as the actual crystal growth rate was used in the simulation.

An on-line control system using the reconstructed model with the minimum order Luenberger observer is illustrated in Figure 7. The dynamic CSD in a continuous crystallizer is calculated by the reconstructed model and the observer H tracks the state variables in the reconstructed model to the real state variables in the real system. The synthesized output from the model and the real system goes into the controller K. With the current technology of fines measurement using light-scattering measurements in a suitably conditioned fines system (Randolph et al., 1986), the control scheme shown in Figure 7 should be feasible.

SUMMARY AND CONCLUSIONS

A numerical algorithm to simulate the dynamics of CSD was developed using the method of lines. The new algorithm demonstrates the numerical stability criteria previously developed for continuous crystallizers. The grid number for numerical calculation could be reduced using the method of lines; computational time is reduced and computer implementation of a simulator becomes feasible. Conventional analytical methods in system theory can be adopted to analyze the crystallizer system because the equations could be cast in a state space form. Nucleation/growth rate kinetics were implemented as an auxiliary function in the algorithm so the main algorithm was independent of kinetics; thus, different kinetics can be easily be substituted. It is possible to define an arbitrary particle size distribution as the initial condition so the algorithm could be utilized for the analysis of any arbitrary growth-type particulate system.

An experimental KCl crystallizer with fines removal was simulated using the new algorithm. Using an arbitrarily determined set of parameters, the crystallizer was properly simulated; the new algorithm could certainly predict dynamic CSD for the idealized R-z complex crystallizer. Finally, a process feasible on-line CSD controller using a minimum order Luenberger observer control algorithm is illustrated for this system.

ACKNOWLEDGEMENT

This work was supported by NSF Grant CBT-811775303.

NOTATION

A = a matrix which is defined by equation (6)
B = an input coefficient matrix
C = an output coefficient matrix
$f_{i,j}$ = fractional moment
$G, \bar{G}$ = dynamic and steady growth rate
h = removal function
i = Nucleation growth rate kinetic sensitivity parameter
j = magma-dependent nucleation kinetics exponent
L = crystal size
$n°, \bar{n}°$ = dynamic and steady nuclei density
Q = a matrix
Qp = mixed production rate
Q(L) = removal rate
R = recycle ratio of the fines dissolver
$\underline{r}$ = an input vector

$u, \bar{u}$	=	dynamic and steady dimensionless number density vector
v	=	vessel volume
x	=	dimensionless particle size
z	=	recycle ratio of the product classifier
β	=	ratio of fines destruction to external production
δ	=	difference of particle size (= ΔX)
σ	=	parameters in integral
θ	=	dimensionless time
τ	=	characteristic time
Φ	=	transition matrix
ϕ	=	dimensionless growth rate

SUBSCRIPTS

F	=	the upper cut size of the fines dissolver
ℓ	=	discrete size notation
m	=	discrete time notation
p	=	the lower cut size of the product classifier

SUPERSCRIPT

'	=	Transpose

LITERATURE CITED

1. Carver, M.B., J.M. Blair, W.N. Selander, and D.G. Stewart, "Short Course and Workshop on Numerical Solution of Ordinary and Partial Differential Equations Using FORSIM", Mathematics and Computation Branch, Chalk River Nuclear Laboratories, Atomic Energy of Canada Limited, Chalk River, Ontario KOJ IJO (1976).

2. Cellier, F., ECE501 Text, Department of Electrical and Computer Engineering, The University of Arizona, Arizona (1985).

3. Kailath, T., "Linear System", Prentice-Hall, New York (1981).

4. Moler, C.B., "MATLAB User's Guide", Department of Computer Science, University of New Mexico, New Mexico (1981).

5. Randolph, A.D., G.L. Beer, and J.P. Keener, "Stability of the Class II Classified Product Crystallizer with Fines Removal", AIChE J., 19, 1140 (1973).

Table 1

Summary of Experimental Data for KCl Crystallizer

Parameters	Experimental Data
x_F	1 [-]
x_p	N/A
R	13 [-]
z	N/A
$\bar{G}$	$\simeq$ 1 [μm/min]
$\bar{n}^\circ$	5.4 x 10^{16} [1/$m^3 \cdot$m]
τ	180 [min]
Qp	100 [ml/min]
V	18 [ℓ]

Table 2

Summary of Parameters Used for Simulation

Parameters	For BRZ=1*	For BRZ=2*
x_F	1	1
x_p	10	6
R	13	13
z	1	5
dQ	0.05	0.05
dx	0.2	0.2

*BRZ represents the semi-log distribution slope between x_F and x_p.

6. Randolph, A.D., L. Chen, and A. Tavana, "Feedback Control of CSD in a KCl Crystallizer with Fines Dissolving", AIChE J. (in press, 1986).

7. Randolph, A.D. and M.A. Larson, "Theory of Particle Process", Academic Press (1971).

8. Tsuruoka, S., "Dynamics and Feedback Control of Crystal Size Distribution in a Continuous Crystallizer", Ph.D. Thesis, the University of Arizona, Tucson, Arizona (1986).

9. Vemuri, V., and W.J. Karplus, "Digital Computer Treatment of Partial Differential Equations", prentice-Hall (1981).

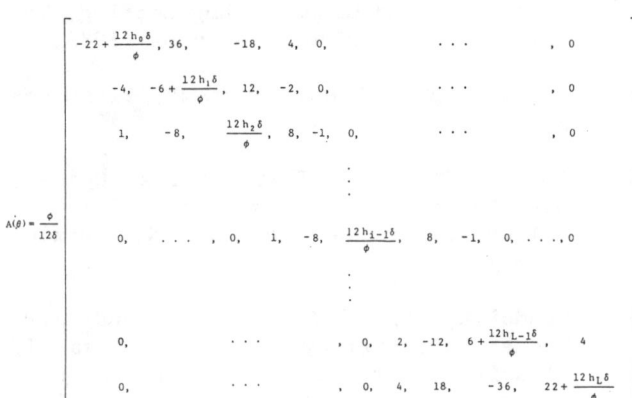

Figure 1. Matrix $A(\theta)$.

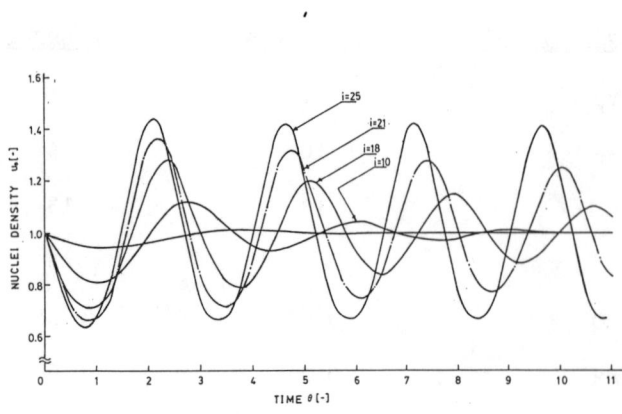

Figure 2. Simulation results for an MSMPR Crystallizer with various nuclei sensitivities, where R=8.5, z=7, x_F=0.2, s_p=3, and j=0.

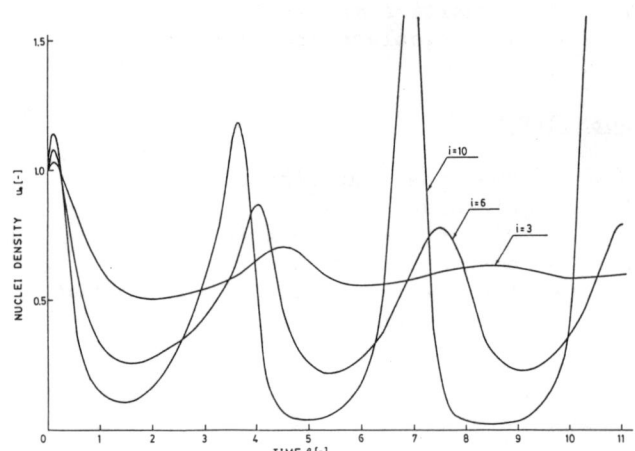

Figure 3. Simulation results in nuclei density for an R-z crystallizer with various nuclei sensitivities, where R=8.5, z=7, x_F=0.2, x_p=3, and j=0.

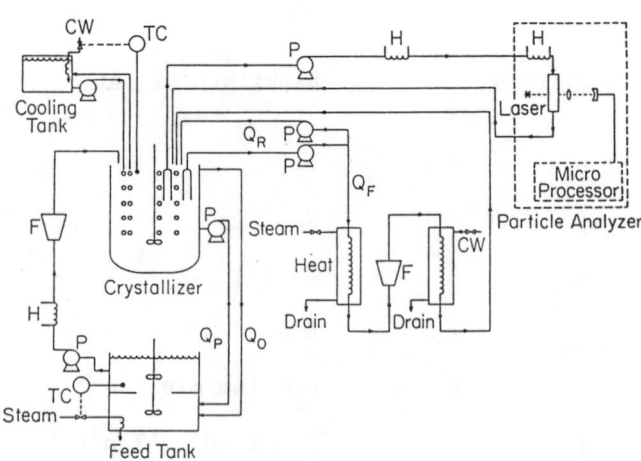

Figure 4. Experimental and theoretical population density at $L=4.58 \times 10^{-4}$[m].

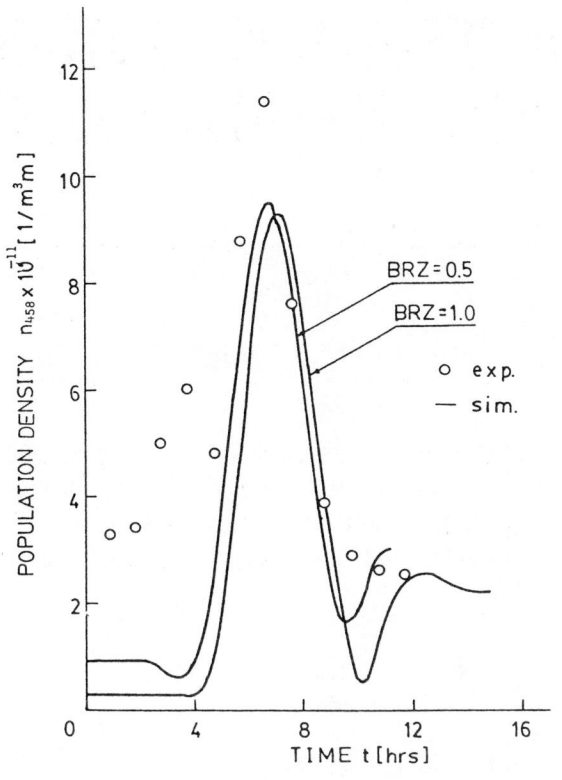

Figure 5a. Experimental and theoretical population density at $L=4.58 \times 10^{-4}$ [m].

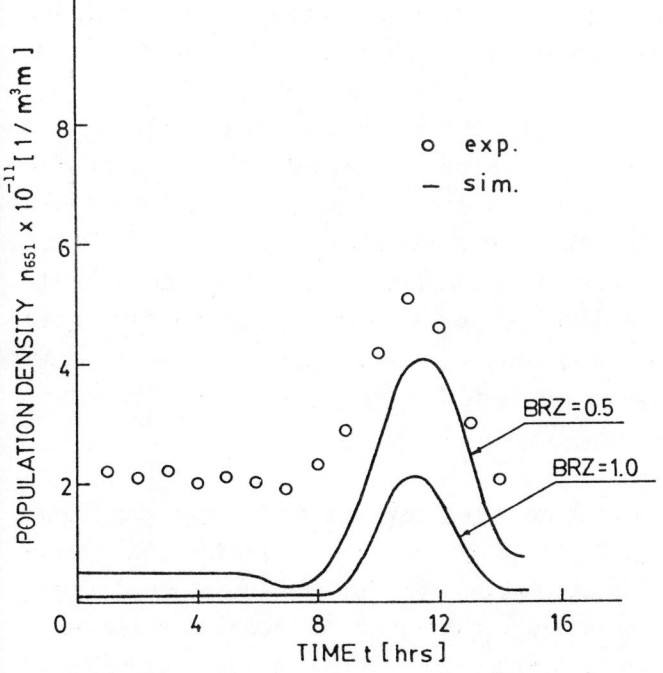

Figure 5b. Experimental and theoretical population density at $L=6.51 \times 10^{-4}$ [m].

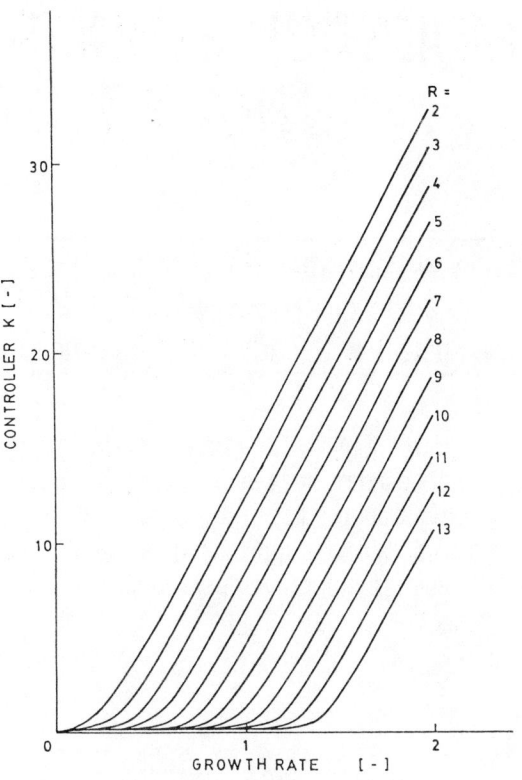

Figure 6. Optimal controllers for various recycle ratios R, where $x_F=1.0$, $x_p=6$, $i=3$, and $j=1$.

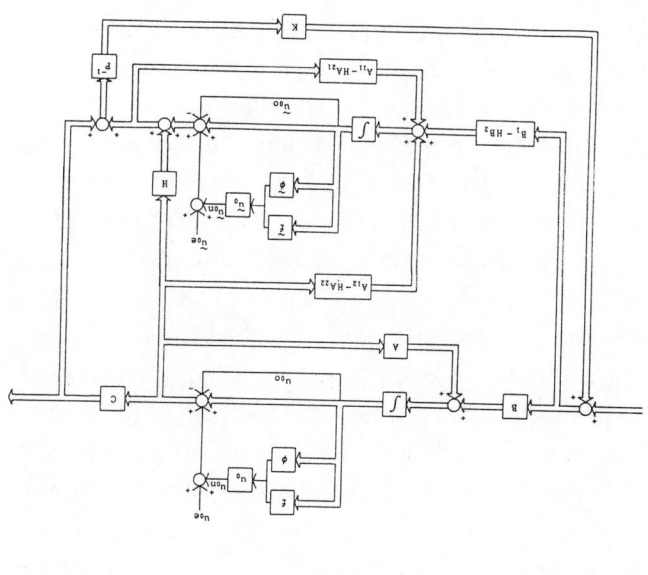

Figure 7. A schematic flow diagram of multiple input-multiple output system with the minimum order Luenberger observer.

SIMULATION STUDIES OF A FEEDBACK CONTROL STRATEGY FOR BATCH CRYSTALLIZERS

Chuei-Tin Chang ■ Department of Chemical Engineering, University of Nebraska, 236 Avery Laboratory, Lincoln, NE 68588-0126
Mary A. Farrell Epstein ■ Department of Pharmocology, University of Connecticut, Farmington, CT 06032

In general, the control objective of a batch crystallization process can be considered as achieving a specific value of a predefined performance index, such as the average product crystal size. A feedback control strategy, based on the idea of one-step iteration, was adopted for this purpose. Both open- and closed-loop simulation studies were carried out to demonstrate the feasibility of this approach. The results show satisfactory and expected improvement when compared with the programmed cooling/evaporative operations.

This study was motivated by a series of publications concerning the control of batch crystallizers with a strategy of programmed cooling/evaporative operation (1,2,3,4,5). In these studies, the sequence of operating conditions, especially the crystallizer temperature, was predicted as a function of time but not as a function of the state variables of interest. Mullin and Nyvlt (1) performed the first study to compute the cooling curve for a batch crystallization operation such that the product average size would be increased. Jones and Mullin (3) carried out a similar investigation with a slightly different set of equations. Jones (4) solved the same problem using a different approach, i.e. the Pontryagin Continuous Maximum Principle (6). He used moment equations, obtained by transformation of the population balance equation, as the system equations to determine the optimal cooling curve for maximizing the average crystal size. The significance of these studies is that these open-loop control policies were carried out experimentally. Thus, programmed operations should be considered practically feasible for batch crystallization processes.

One natural extension of the above studies is to implement a certain form of closed-loop control, because the feedback controller is responsive to the instantaneous perturbations during operation. Since the control objective for a batch crystallization process can be achievement of a specific value of a predefined performance index, e.g. the average crystal size, the candidate feedback control strategy should possess the capability of maintaining the desired performance index value despite the unexpected disturbances in the state variables.

A new strategy has been developed based on the idea of one-step iteration (7). Under the assumption that the perturbed trajectory is in the vicinity of the nominal trajectory, this strategy improves the perturbed performance index in a manner similar to a gradient search technique. In this paper, a summary of the derivation and the results of the corresponding numerical simulations are presented to demonstrate the feasibility of the proposed control policy.

THE MATHEMATICAL MODEL OF A BATCH CRYSTALLIZATION PROCESS

The lumped parameter system equations for batch crystallization processes, obtained by moment transformation of the population balance equation, describe the dynamic behavior of the leading moments of Crystal Size Distribution or CSD (8). For a perfectly mixed batch crystallizer with a solute-solvent system that exhibits size-independent crystal growth rate, the system governing equations can be expressed as

$$\frac{dm_0}{dt} = B^O G(m_3,T)$$

$$\frac{dm_1}{dt} = m_0 G(m_3,T)$$

$$\frac{dm_2}{dt} = 2m_1 G(m_3,T)$$

$$\frac{dm_3}{dt} = 3m_2 G(m_3,T)$$

(1)

Here, B^O is the nucleation rate, G is the crystal growth rate, m_0, m_1, m_2, and m_3 represent the zeroth, first, second and third moments of CSD respectively. The solute-solvent system used in this investigation is KNO_3-water. The growth and nucleation rate expressions and the kinetic data are taken from the work of Helt and Larson (9). They are:

$$G = 4.8 \times 10^6 \exp\{-31/[R(T+273)]\} \Delta C \quad (2)$$

$$B^O/M_T = 3.0 \times 10^{-10} \exp\{108/[R(T+273)]\} \Delta C^{1.7} \quad (3)$$

where ΔC is the supersaturation of solute in solvent and M_T is the total mass of crystals in the crystallizer.

The control strategy described in this paper involves varying the manipulated variable, i.e. temperature, to maintain a specified value of a performance index. The three performance indices used in the simulation studies are

1) the number average size,
$$J_1 = (m_1/m_0)_{t=t_f},$$

2) the total volume,
$$J_2 = m_3]_{t=t_f}, \text{ and}$$

3) the CSD variance,
$$J_3 = [(m_2/m_1) - (m_1/m_0)^2]_{t=t_f},$$

of the product crystals. The variable t_f denotes the ending time of the batch crystallization.

THE FEEDBACK CONTROL STRATEGY

For illustration purpose, the problem described in the previous section is written in a general formulation. The performance index can be expressed as

$$J = K[x(t_f), t_f] \quad (4)$$

where K is an arbitrary function of $x(t_f)$ and t_f, and $x(t)$ represents the trajectories of the state variables governed by the following system equations

$$\dot{x} = f(x,u,t) \quad \text{and} \quad x(t_0) = x_0 \quad (5)$$

In equation (5), u represents the manipulated variables and t_0 represents the starting time for the crystallization. When the nominal trajectories of the manipulated variables, $u_n(t)$, are applied to the system, the resulting trajectories of the state variables, $x_n(t)$, can be determined by equations (4) and (5) respectively. Note, that these nominal trajectories do not have to be optimal, although optimal trajectories are not excluded from this category.

If small perturbations from the nominal paths are introduced in the initial state, i.e. $\delta x(t_0)$, one can expect that there are also perturbations in the state variables during operation, i.e. $\delta x(t)$ for $t_0 < t < t_f$. The difference between the resulting performance index, J, and the nominal performance index, J_n, can be expressed as

$$\Delta J = J - J_n = \delta J + \delta^2 J + \ldots \quad (6)$$

Since $x_n(t)$ is not required to be optimal, the first variation of J may or may not vanish on the nominal trajectories. In order to achieve our goal, i.e. keeping the absolute value of ΔJ as small as possible, let's define a new performance index,

$$I = (\Delta J)^2 \quad (7)$$

If this new performance index is minimized, then, our orginal control objective is also accomplished.

Under the assumption that the first variation of the original performance index, J, in equation (6) dominates, the new performance index, I, can be approximated by another performance index, $\hat{I}$, and

$$\hat{I} = (\delta J)^2$$
$$= (X^T S_f X)/2 \,]_{t=t_f} \quad (8)$$

where, $X = \delta x$ and $S_f = 2 K_x K_x^T]_{t_f}$.

The deviations of state variables from the nominal trajectories can be described by the linearized system equations:

$$\dot{X} = A(t)X + B(t)U \quad \text{and} \quad X(t_0) = X_0 \quad (9)$$

where, $U = \delta u$, $A = f_x$ and $B = F_u$.

Generally speaking, a feedback control policy is concerned with the problem of assigning the value of U as a function of X. This problem is very similar to that of a numerical iteration process for determining a minimum (or maximum) performance index. If the trajectory, $U(t) = \delta u(t) = 0$ for $t_0 < t < t_f$, is considered as the "initial guess" in an numerical iteration process, the improved U(t) in the next iteration can be obtained based on the "initial" non-optimal trajectories of the state variables.

Starting with equations (8) and (9), subject to the initial guess $U^{(0)}(t) = 0$, the "uncontrolled" trajectory, $X^{(0)}(t)$ for $t_0 < t < t_f$, can be predicted. Based on this trajectory, the improved manipulated variable, $U^{(1)}$, can be obtained using the well-konown Gradient method (9). The resulting expression for the improved manipulated variable is presented below:

$$U^{(1)} = -\varepsilon B^T \lambda^{(0)} \quad (10)$$

where ε is the step size and $\lambda^{(0)}$ satisfies the following equations:

$$\dot{\lambda}^{(0)} = -A^T \lambda^{(0)}, \quad \lambda^{(0)}(t_f) = S_f X^{(0)}(t_f) \quad (11)$$

and

$$\dot{X}^{(0)} = AX^{(0)}, \quad X^{(0)}(t_0) = X_0 \quad (12)$$

In order to achieve feedback, let's assign

$$\lambda^{(0)}(t) = S(t)X^{(0)}(t) \quad (13)$$

and, then, substitute equation (13) into equation (10). The resulting feedback control strategy can be written as

$$U^{(1)} = \varepsilon \Lambda X^{(0)} \quad (14)$$

where, $\Lambda = -B^T S$ and ε can be considered as an adjustable parameter of the feedback controller.

The trajectory of B(t) can be obtained using the definition presented in equation (9). Also, it has been shown (7) that the trajectory of S(t) can be determined by

$$\dot{S} = -SA - A^T S \text{ and } S(t_f) = S_f \quad (15)$$

A block diagram of the control system is presented in Figure 1. It can be observed that on-line storage of the trajectories of the feedback multipliers, $\Lambda(t)$, the nominal state variables, $x_n(t)$, and manipulated variables, $u_n(t)$, is necessary. Since the results of this derivation are independent of the initial conditions, these trajectories can be calculated before actual implementation of the proposed strategy.

INITIAL CONDITIONS AND NOMINAL TEMPERATURE PROFILES USED IN NUMERICAL SIMULATION STUDIES

The seeds chosen for the simulation studies are uniformly distributed with a constant population density, $n=10^5$ number of particles/cm/g H_2O, over a size range from 2.5×10^{-3} to 22.5×10^{-3} centimeters. Thus, the first four moments of the seed distribution have the following values: $m_0=2000.0$, $m_1=25.0$, $m_2=0.3792$, $m_3=0.006406$. They will be used as the initial conditions of the system equations (1) throughout the simulation studies. Also, the initial solute concentration is 1.15 g KNO_3/g H_2O for all the cases. The dynamic behavior of the concentration during operation can be determined by material balance.

Two linear temperature profiles for a period of 1000 seconds are arbitrarily chosen as the nominal trajectories of the manipulated variable. The initial temperatures are the same for both profiles, $55^\circ C$. The final temperatures are 45 and $40^\circ C$ respectively. They will be referred as Trajectories I and II later in this paper. Neither selected trajectory optimizes the performance index defined in equation (4). The corresponding nominal trajectories of the state variables, i.e. $m_0(t)$, $m_1(t)$, $m_2(t)$ and $m_3(t)$, were generated by numerically integrating equation (1).

OPEN-LOOP SIMULATION

In deriving the feedback control strategy for the batch crystallization processes, two questions naturally arise: 1) how well can the square of the first variation, $(\delta J)^2$, approximate the new performance index, $I=(\Delta J)^2$, and, especially, 2) how well can the results of optimizing the quadratic performance index, equation (8), with linear system equations, equation (9), guide us to improve the performance index in equation (4) with non-linear system equations, equation (5).

In this part of our study, we sought to answer these questions by open-loop numerical simulations. First, One-step iteration calculations were carried out for the transformed optimal (linear) control problem, equations (8) and (9), to determine the temperature deviation profiles, $\delta u(t)$, needed for decreasing the value of the performance index, $\hat{I}$. Then, open-loop simulations were performed using equation (5), or equation (1), with and without the temperature deviation profile.

Three sets of studies were carried out. Each corresponds to a performance index defined in the model section, i.e. J_1, J_2 and J_3. For each set of studies, two nominal trajectories were generated using the two nominal temperature profiles selected earlier. To create perturbations, disturbance factors, F, were introduced in the initial state variables, i.e. $x(t_0) = F x_n(t_0)$, where $0 < F < 2$ and $F \neq 1$. Since the secondary nucleation always takes place, the magnitude of disturbances in the initial state is small when compared with the values of the state variables at the end of operation. Thus, the previously mentioned approximations are generally valid in the simulation studies.

Results corresponding to J_1 (the number average size) and J_3 (the CSD variance) are provided in this paper. The results corresponding to J_2 were reported elsewhere (<u>7</u>). Since the calculated temperature deviation profiles are similar for both nominal trajectories, only two for each performance index are plotted as examples in Figures 2 and 3. One is the temperature deviation profile required with positive initial disturbance (F>1) introduced, the other is the temperature deviation profile required with negative initial disturbance (F<1) introduced.

Based on equations (8) and (9), the temperature deviation profile represents the gradient needed for reducing the value of a given performance index in a numerical iteration process. Perfectly reasonable physical interpretations of the gradient can be obtained by investigating the results of open-loop simulations of the original non-linear system equations with and without the calcuated temperature deviation profile. Those results are presented in Tables IA and IB. The definitions of the symbols used in these tables are provided in the Notation section at the end of this paper.

From Table IA, it can be observed that the positive disturbances cause an increase in the product average size (E1>0). From the corresponding temperature deviation profile, Figure 2a, and equations (2) and (3), it can be observed that the temperature variation at the beginning of the operation tends to slow the growth of the seeds initially and then creates more nuclei after the seeds have already grown for some time. At the end of operation, the sudden drop in temperature will undoubtedly generate a burst of tiny nuclei. Both actions can reduce the final number average size so that the perturbed performance index will be maintained at its nominal value. The required temperature deviations for cases with negative disturbances are the opposite of those just described (see Figure 2b) and can be interpreted by the same reasoning.

Tabel IB shows us that the product CSD variance is decreased (E1<0) for positive disturbances. It is important to recall that the objective of the control strategy is to maintain the performance index at its nominal level. The temperature peak with respect to the nominal profile in Figure 3a will certainly reduce the high supersaturation that appears at the beginning of crystallization. This will flatten the final CSD at the size where maiximum number density is located. Furthermore, the final drop in temperature will generate a large number of small crystals, which will also increase the resulting variance of the product CSD in accordance with the goal of the control strategy. A similar interpretation can be obtained for the temperature deviation profile presented in Figure 3b with negative disturbance. Thus, it can be concluded that study of the gradient used in numerical iteration procedure can provide revealing insights into batch crystallizer operation.

A comparison between the $\hat{I}$'s and the $(\Delta J)^2$'s is presented in Tables IA and IB. It can be observed that the ratio, $RT=\hat{I}/(\Delta J)^2$, has a value ranging from 3.5 to 51.2. However, the desired values of the performance indices can still be obtained by varying the step size, ϵ. Examples of the effect of changing step size are also presented in Tables IA and IB. In each example, the approximate range of the step size value for achieving the desired performance index is indicated by an arrow. Thus, although the transformed $\hat{I}$ itself is not a good approximation to $(\Delta J)^2$, the predicted correction does lead us in a right direction.

CLOSED-LOOP SIMULATION

The feedback control of batch crystallization process can be realized by the calcuation and on-line storage of the trajectories of the feedback multipliers, $\Lambda(t)$, defined is equation (14). The closed-loop simulations are programmed according to the arrangement presented in Figure 1. Although open-loop simulations of equation (1) with adjusted temperature profile indicate the corrective measure is successful, it is still necessary to test the applicability of the feedback control scheme using the time dependent gain $\Lambda(t)$. This is because the feedback scheme allows the system to modify its manipulated variable continously, but the improved open-loop simulations make only one adjustment.

To be more specific, let's compare the difference between the two in detail. In both cases, the system governing equation can be generalized by the following expression:

$$\dot{x} = f(x, u_n + \delta u, t) \quad (16)$$

For the open-loop simulation, the temperature deviation, $\delta u(t)$, is calculated by the following equation:

$$\delta u(t) = -\epsilon B^T S \, \delta x(t) \quad (17)$$

where $\delta x(t)$ is the perturbed state variable estimated by equation (12). Once the values of $\delta x(t)$'s for $t_0 < t < t_f$ are obtained, the values of $\delta u(t)$'s can be determined for every time increment before the simulation of equation (16).

In the case of closed-loop simulation, the deviations in state variables at each time increment are calculated by:

$$\delta x(t) = x(t) - x_n(t) \quad (18)$$

where $x_n(t)$ is the nominal state trajectory obtained by simulation of equation (16) with $\delta u(t)=0$ for $t_0 < t < t_f$. At every time increment of the closed-loop simulation, the temperature deviation is calculated by equation (17) and used in equation (16). From equations (12), (16), (17) and (18), it is obvious that different trajctories, $x(t)$, and different performance index values will be generated for open- and closed-loop control system simulations with the same disturbance factor and step size.

The results of the closed-loop simulations of the corresponding cases in Tables IA and IB are presented in Tables IIA and IIB. The definitions of the symbols used in these tables can be found in the Notation section. Examples showing the effect of varying the step size is also presented in these tables. From Tables IIA and IIB, it can be observed that all the performance index values resulting from closed-loop control simulations are much closer to the nominal values than the ones without control. Comparing the cases in Tables IA and IB with the ones in Tables IIA and IIB, it can be observed that the values used in closed-loop simulations are larger than those used in the corresponding open-loop cases, however, the desired nominal value of the performance index can still be achieved (or approached).

CONCLUSION

A feedback control strategy for batch crystallization processes that can be used to maintain a performance index at a specific nominal value has been presented in this paper. The concept of one-step iteration was adopted to achieve this control objective. Various numerical simulations, both open- and closed-loop, were carried out to demonstrated the feasibility of the proposed strategy. The results show satisfactory and expected improvement. In addition, since the formulation of this control policy is rather general, it can also be applied to other batch systems, such as the batch chemical reactor, as well.

NOTATION:

B^o = nucleation rate, number of particles/sec/g H_2O
C = concentration, g solute/g H_2O
ΔC = supersaturation, g solute/g H_2O
f = system governing equations
$E1 = (J-J_n)/J_n$, %
$E2 = (J_{imp}-J_n)/J_n$, %
F = disturbance factor
G = growth rate, cm/sec
I = performance index, $I=(\Delta J)^2$
$\hat{I}$ = approximation of I, $\hat{I}=(\delta J)^2$
J = performance index
K = expression of performance index
m_0 = the zeroth moment of CSD, number of particles/g H_2O
m_1 = the first moment of CSD, cm/g H_2O
m_2 = the second moment of CSD, cm^2/g H_2O
m_3 = the third moment of CSD, cm^3/g H_2O
M_T = the total mass of crystals in crystallizer, g solids/g H_2O
$RT = \hat{I}/(\Delta J)^2$
$SS = \epsilon/10^7$
t = time, sec
T = temperature, ^{o}C
x = state variable
X = deviation of state variable from its nominal value
u = manipulated variable
U = deviation of manipulated variable from its nominal value

Greek Letters

λ = adjoint variable
ϵ = step size
Λ = time dependent feedback multiplier

Superscript

(0) = initial guess
(1) = improved by one-step iteration

Subscript

n   = nominal value
0   = initial condition
f   = final condition
imp = improved value

LITERATURE CITED:

1. Mullin, J. W. and J. Nyvlt, *Chem. Eng. Sci.*, 26, 369 (1971).

2. Larson, M. A. and J. Garside, *The Chemical Engineer*, 318, (June 1973).

3. Jones, A. G. and J. W. Mullin, *Chem. Eng. Sci.*, 29, 105 (1974).

4. Jones, A. G., *Chem. Eng. Sci.*, 29, 1075 (1974).

5. Tavare, N. S. and M. R. Chivate, *The Chem. Eng. J.*, 14, 175 (1977).

6. Pontryagin, L. S., et al, *The Mathematical Theory of Optimal Processes*, English Edition, Interscience, New York (1962).

7. Chang, C. T., "Control Strategies for Batch Crystallization Processes," PhD Thesis, Columbia University, New York (1982).

8. Randolph, A. D. and M. A. Larson, *Theory of Particulate Processes*, Academic Press, New York (1971).

9. Kelly, H. J. "Method of Gradients," in *Optimization Techniques*, Leitman, G. (Ed.), Academic Press, New York (1962)

Table IA Results of Open-loop Simulations for J_1

Trajectory I			Trajectory II		
$J_n = 0.03129$			$J_n = 0.02654$		
F	E1	RT	F	E1	RT
1.3	+1.89	10.6	1.2	+2.75	4.0
1.2	+1.34	8.8	1.1	+1.43	3.8
1.1	+0.70	7.5	1.05	+0.72	3.7
0.9	-0.93	5.6	0.95	-0.72	3.5
0.8	-1.95	4.8	0.9	-1.47	3.5
0.7	-3.10	4.2			

Effect of varying step size:

F = 1.3		F = 1.2	
SS	E2	SS	E2
0.00	+1.89	0.0	+2.75
0.07	+0.99	0.1	-0.83 ←
0.10	+0.54		
0.50	-7.89 ←		

Table IB Results of Open-loop Simulations for J_3

Trajectory I			Trajectory II		
$J_n = 0.3849 \times 10^{-3}$			$J_n = 0.4789 \times 10^{-3}$		
F	E1	RT	F	E1	RT
1.3	-11.22	10.9	1.3	-5.74	12.1
1.2	-7.46	11.0	1.2	-3.61	13.6
1.1	-3.69	11.3	1.1	-1.67	15.6
0.9	+3.51	12.6	0.9	+1.38	23.6
0.8	+6.70	13.8	0.8	+2.34	32.9
0.7	+9.38	15.7	0.7	+2.82	51.2

Effect of varying step size:

F = 0.9		F = 1.3	
SS	E2	SS	E2
0.0	+3.51	0.0	-5.74
10.0	+3.07	100.0	+0.77 ←
50.0	+1.33		
70.0	+0.52 ←		

Table IIA Results of Closed-loop Simulations for J_1

Trajectory I			Trajectory II		
$J_n = 0.03129$			$J_n = 0.02654$		
F	SS	E2	F	SS	E2
1.3	1.0	+0.03	1.2	0.4	+0.72
1.2	1.5	−0.13	1.1	1.5	+0.00
1.1	0.6	+0.26	1.05	0.7	+0.11
0.9	2.0	−0.10	0.95	0.7	−0.19
0.8	0.25	−1.50	0.9	0.2	−1.06
0.7	0.125	−2.78			

Effect of varying step size:

F = 1.3		F = 1.2	
SS	E2	SS	E2
0.0	+1.89	0.0	+2.74
0.1	+1.18	0.1	+2.11
0.5	+0.58	0.2	+1.51
1.0	+0.10 ←	0.4	+0.72 ←
1.5	−0.19		

Table IIB Results of Closed-loop Simulations for J_3

Trajectory I			Trajectory II		
$J_n = 0.3849 \times 10^{-3}$			$J_n = 0.4789 \times 10^{-3}$		
F	SS	E2	F	SS	E2
1.3	1000	−1.79	1.3	800	+0.10
1.2	2000	−0.36	1.2	500	−0.13
1.1	2000	−0.03	1.1	400	−0.04
0.9	600	−0.13	0.9	300	0.00
0.8	500	−0.42	0.8	300	−0.42
0.7	500	−1.33	0.7	200	−0.29

Effect of varying step size:

F = 0.9		F = 1.3	
SS	E2	SS	E2
0.0	+3.51	0.0	−5.74
70.0	+2.57	100.0	−3.97
100.0	+2.26	500.0	−0.69
300.0	+0.73	800.0	+0.10 ←
500.0	+0.05 ←		
600.0	−0.13		

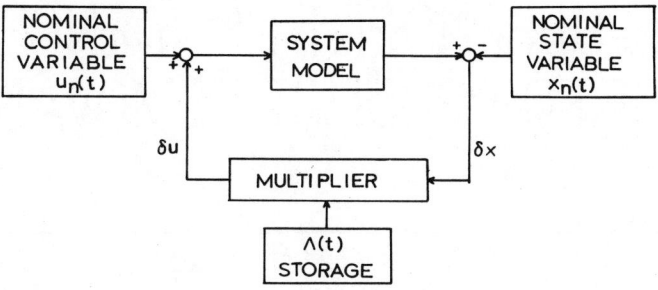

Figure 1. Block diagram of the proposed feedback control system.

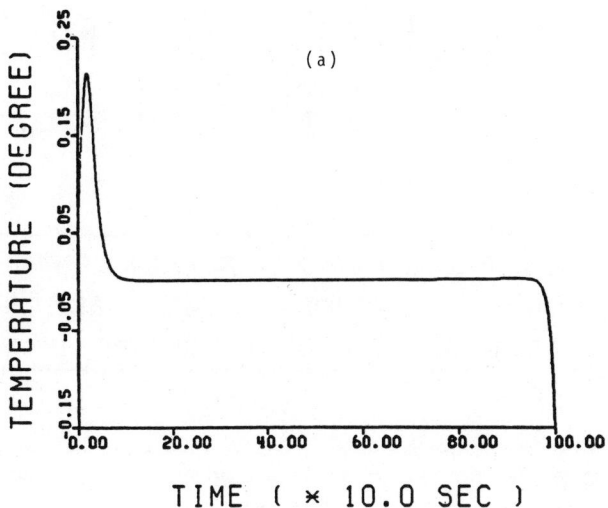

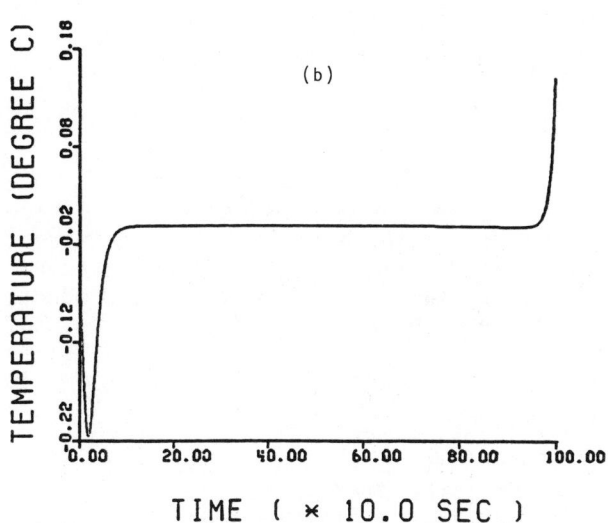

Figure 3. Temperature deviation profiles for J_3
a) $F > 1$; b) $F < 1$.

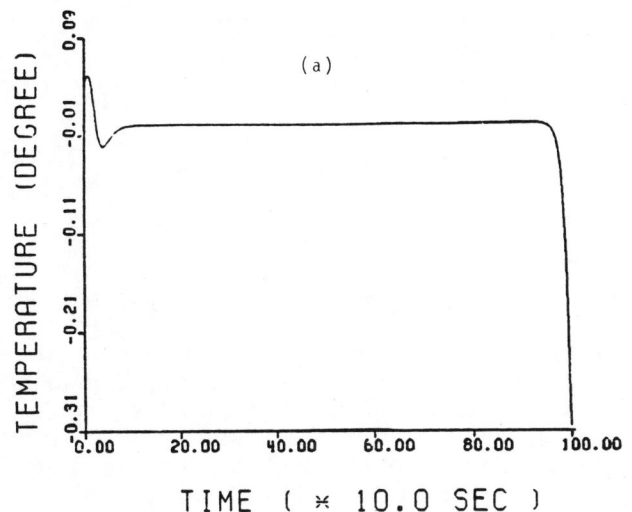

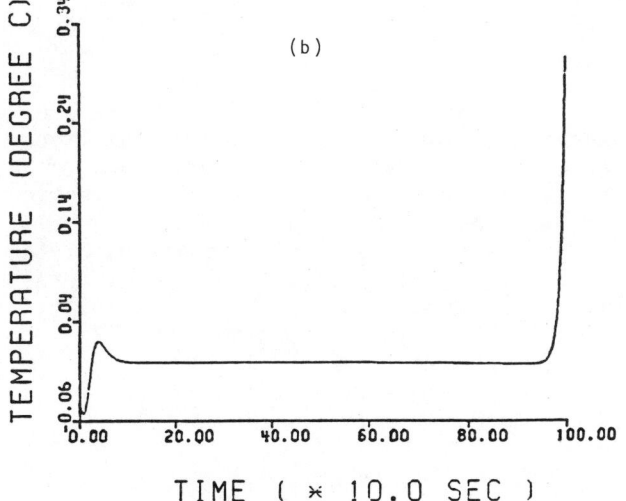

Figure 2. Temperature deviation profiles for J_1
a) $F > 1$; b) $F < 1$.

TRANSIENT ANALYSIS OF CRYSTALLIZATION: EFFECT OF THE SIZE-DEPENDENT RESIDENCE TIME OR CLASSIFIED PRODUCT REMOVAL

L. T. Fan ■ Department of Chemical Engineering, Kansas State University, Manhattan, KS 66506
S. T. Chou ■ Department of Statistics, Kansas State University, Manhattan, KS 66506
J. P. Hsu ■ Department of Chemical Engineering, National Taiwan University, Taipei, Taiwan, R.O.C.

Effect of the size-dependent residence time distribution or classified product removal on the transient behavior of a continuous mixed suspension, mixed product removal crystallizer has been investigated by modeling and simulating it stochastically. It has been shown that a linear increase in the residence time with the increase in crystal size results in an upward curvature in its distribution at its lower end; if the residence time of a crystal is inversely proportional to its size, a downward curvature at the lower end of its distribution is predicted.

Crystals of a predtermined size range often are removed preferentially from a continuous mixed suspension, mixed product removal (MSMPR) crystallizer as the desired product; this renders the residence time of a crystal to depend on its size (1). This size-dependent residence time can also be due to the hydrodynamic conditions inside the crystallizer. For example, rather than circulated with the liquid phase, relatively large crystals tend to accumulate at the bottom of the crystallizer and be removed with slurry as a product. In contrast, relatively small crystals tend to be entrained in the upflow liquor and exit from the crystallizer with the overflow.

The objectives of the present study are to investigate the influence of size-dependent residence time of a crystal on the transient crystal size distribution (CSD), and to examine if the size-dependent residence time can be a major cause for the behavior of the crystallizer to deviate appreciably from that of an idealized crystallizer (2). An MSMPR crystallizer with a constant growth rate of crystals but without crystal breakage and aggregation is regarded as an idealized crystallizer. Furthermore, it is assumed that the idealized crystallizer is uniformaly mixed, and thus the crystals are assumed to have the same residence-time distribution as those of the liquid components.

Crystal growth is often accomplished by the transport of fluctuating numbers of solute molecules through a turbulent flow field to the surfaces of growing crystals which are mesoscopic particulate entities. Complicated interactions among the solute molecules in the liquid phase and the mesoscopic particles in the solid phase are expected to be chaotic and random. Thus, the phenomenon of crystallization can not be portrayed precisely by a deterministic model; naturally, it is highly desirable that a stochastic approach need be adopted (3, 4).

Although knowledge of the transient behavior of the CSD is of fundamental importance for the start-up and control of a crystallizer, relatively little attention has been paid to its analysis (5, 6, 7). The deterministic governing equations of the crystallizer are usually partial differential equations, and solving or simulating them is rather time-consuming.

The steady-state CSD, $n(L)$, in an idealized MSMPR crystallizer is (2)

$$n(L) = n^o \exp(-L/G\tau) \qquad (1)$$

This expression suggests that a semi-logarithmic plot of $n(L)$ against the crystal size, L, is linear. However, such a plot is often found to exhibit an upward curvature at small crystal sizes. Deviations of the behavior of a crystallizer from that of the idealized crystallizer have been attributed to various factors, including the size-dependent growth rate, growth rate dispersion, initial-size distribution, and classified product removal (1, 2, 8, 9, 10, 11, 12, 13). It should be pointed out that such a deviation could be caused by one or more of these factors. Thus, each factor need be examined separately and exhaustively as a possible cause.

MODELING

Consider an isothermal MSMPR crystallizer in which seeds can appear along with the input stream or by nucleation. Each crystal stays a random period of time inside the crystallizer. During this period, it may grow in size or maintain its original size. The crystal size is partitioned into integral multiples of increment δ; a crystal in state i has a size of $i\delta$; and a seed is in state 1. The crystal growth is portrayed as a process of attachment of a cluster of solute molecules of size δ to the crystal surface.

A crystal is considered not to break up during its growth, and the probability of simultaneous attachment of two or more clusters to the crystal surface in a small time interval is negligible. Thus, if a crystal leaves state i in a small time interval due to growth, it can only proceed to the next higher state, state (i+1). In deriving the CSD of the present crystallizer we shall further assume the following.

1. The conditional probability that a crystal will be in state (i+1) at time $(t+\Delta t)$, given that it is in state i at time t, is $[\lambda_i \Delta t + o(\Delta t)]$ where λ_i is the transition intensity at state i.

2. The conditional probability that a crystal will exit from the crystallizer at time $(t+\Delta t)$, given that it is in state i at time t, is $[\mu_i \Delta t + o(\Delta t)]$ where μ_i is the intensity of exit at state i.

3. The conditional probability that a crystal will stay in state i at time $(t+\Delta t)$, given that it is in state i at time t, is $[1-\lambda_i \Delta t - \mu_i \Delta t + o(\Delta t)]$ where $o(\Delta t)$ is such that

$$\lim_{\Delta t \to 0} \frac{o(\Delta t)}{\Delta t} = 0$$

The transition diagram of the present model is illustrated in Figure 1.

Master Equation Derivation

Let $p_i(t)$ be the absolute probability or simply probability that a crystal is in state i at time t. By assumption, a crystal in state i, i = 1,2,..., at time t will be in one of the three states, state (i+1), exit state, and state i, with probabilities $[\lambda_i \Delta t + o(\Delta t)]$, $[\mu_i \Delta t + o(\Delta t)]$ and $[1-(\lambda_i+\mu_i)\Delta t + o(\Delta t)]$, respectively, at time $(t+\Delta t)$, where Δt is an infinitesimally small but finite time interval. Thus, the probability that a crystal is in state i at time $t+\Delta t$ is

$$p_i(t+\Delta t) = [\lambda_{i-1}\Delta t + o(\Delta t)]p_{i-1}(t)$$
$$+ [1-(\lambda_i+\mu_i)\Delta t + o(\Delta t)]p_i(t) \quad (2)$$

After rearrangement and division by Δt, Equation (2) becomes

$$\frac{p_i(t+\Delta t)-p_i(t)}{\Delta t}$$
$$= \lambda_{i-1}p_{i-1}(t)-(\lambda_i+\mu_i)p_i(t) + \frac{o(\Delta t)}{\Delta t} \quad (3)$$

Taking the limit as $\Delta t \to 0$, we obtain

$$\frac{dp_i(t)}{dt} = \lambda_{i-1}p_{i-1}(t)-(\lambda_i+\mu_i)p_i(t), \quad i=2,3,... \quad (4)$$

The balance of probability for i=1 yields

$$p_1(t+\Delta t) = [1-(\lambda_1+\mu_1)\Delta t + o(\Delta t)]p_1(t) \quad (5)$$

Following the same procedure as that employed for i=2,3,..., we obtain

$$\frac{dp_1(t)}{dt} = -(\lambda_1+\mu_1)p_1(t) \quad (6)$$

This equation in conjunction with Equation (4) comprises the master equation of the crystallizer under consideration.

Solution of the Master Equation

Equation (4) can be solved subject to the initial condition $p_1(0)=1$; the resultant expression is

$$p_1(t) = e^{-(\lambda_1+\mu_1)t} \quad (7)$$

Substituting Equation (7) into Equation (4) with i=2 and rearranging the resultant expression yield

$$\frac{dp_2(t)}{dt} + (\lambda_2+\mu_2)p_2(t) = \lambda_1 p_1(t)$$
$$= \lambda_1 e^{-(\lambda_1+\mu_1)t} \quad (8)$$

The solution of Equation (8), subject to the initial condition $p_2(0)=0$, is

$$p_2(t) = \frac{\lambda_1}{(\lambda_2+\mu_2)-(\lambda_1+\mu_1)}$$
$$\cdot [e^{-(\lambda_1+\mu_1)t} - e^{-(\lambda_2+\mu_2)t}] \quad (9)$$

Similarly, the solution of Equation (4) with i=3,4,..., subject to the initial conditions $p_i(0)=0$, i=3,4,..., yields

$$p_i(t) = \sum_{k=1}^{i} \left\{ \left[\frac{\prod_{\ell=1}^{i-1} \lambda_\ell}{\prod_{\substack{\ell=1 \\ \ell \neq k}}^{i} [(\lambda_\ell+\mu_\ell)-(\lambda_k+\mu_k)]} \right] e^{-(\lambda_k+\mu_k)t} \right\},$$
$$i=3,4,... \quad (10)$$

In what follows, the transition intensities among states will be considered to be equal and constant, i.e.,

$$\lambda_1 = \lambda_2 = \ldots = \lambda$$

This is equivalent to considering that the growth rate is state or size-independent. In this case, Equations (7), (9) and (10) can be combined as

$$p_i(t) = \begin{cases} e^{-(\lambda+\mu_1)t} & , i=1 \\ \sum_{k=1}^{i} \left\{ \left[\frac{(\lambda)^{i-1}}{\prod_{\substack{\ell=1 \\ \ell \neq k}}^{i} (\mu_\ell-\mu_k)} \right] e^{-(\lambda+\mu_k)t} \right\} & , i=2,3,... \end{cases} \quad (11)$$

Mean and Variance of Number of Crystals of an Individual Size

Let $m(0)$ be the number of seeds initially in the crystallizer. At any moment s, seeds can appear in the crystallizer either through nucleation or with the feed stream. By denoting the rate of appearance of the seeds at time s (s>0) as $z(s)$, the number of seeds appearing in the crystallizer during the time interval (s,s+ds) is $z(s)ds$. These seeds must be in one of the states i, i=1,2,..., in the crystallizer or exit from it at time t (t>s). If it can be assumed that the growth of an individual crystal is independent of each other, the number distribution of crystals in the crystallizer is multinomial with parameters [$z(s)ds$, $p_1(t-s)$, $p_2(t-s)$, ...], and the mean or expected value of the marginal distribution is (14).

$$E[N_i(t-s)] = p_i(t-s)z(s)ds \quad (12)$$

and its variance is (14)

$$Var[N_i(t-s)] = p_i(t-s)[1-p_i(t-s)]z(s)ds \quad (13)$$

where $N_i(t-s)$ is the random variable representing the number of crystals in state i in the crystallizer at time t, which have come from $z(s)ds$ seeds. Let $U_i(t)$ be the random variable representing the total number of crystals in state i accumulated inside the crystallizer up to time t, the mean and variance of $U_i(t)$ are, respectively,

$$E[U_i(t)] = \int_0^t z(s)p_i(t-s)ds + m(0)p_i(t) \quad (14)$$

and

$$\text{Var}[U_i(t)] = \int_0^t z(s)p_i(t-s)[1-p_i(t-s)]ds$$
$$+ m(0)p_i(t)[1-p_i(t)] \qquad (15)$$

Note that the seeds appear in the crystallizer in a continuous manner, and therefore, Equations (12) and (13) need be integrated with respect to time to obtain Equations (14) and (15), respectively. Thus, the mean and variance of $U_i(t)$ can be evaluated by substituting Equation (11) into Equations (14) and (15), respectively. Higher moments of $U_i(t)$ can be evaluated in a similar manner. Also note that the moments here refer to those of the random variables U_i's and are different from moments defined in the deterministic population balance models (see APPENDIX A).

NUMERICAL SIMULATION

The characteristics of the present model are examined through numerical simulation. Two types of size-dependent exit intensity functions are considered for illustration, the first of which takes the form

$$\mu_i = c_1 i \qquad (16)$$

where c_1 is a constant. Equation (16) which is newly proposed here implies that the exit transition intensity is proportional to the size of a crystal, and therefore, the bigger the crystal the larger the probability of it exiting from the crystallizer or the smaller the tendency of it to reside in the crystallizer. For this case, Equation (11) reduces to

$$p_i(t) = \begin{cases} e^{-(\lambda+c_1)t} & , i=1 \\ \sum_{k=1}^{i}\left\{\left[\dfrac{(\lambda)^{i-1}}{\prod\limits_{\substack{\ell=1\\\ell\neq k}}^{i} c_1(\ell-k)}\right]e^{-(\lambda+c_1 k)t}\right\} & , i=2,3,\ldots \end{cases} \qquad (17)$$

The mean CSD is evaluated by substituting Equation (17) into Equation (14). For simplicity, it is assumed that the rate of seeds appearing, $z(s)$, is numerically equal to the number of seeds initially present in the crystallizer, $m(0)$. The transient mean CSD calculated for $C_1=0.1$ is illustrated in Figure 2. The variance around the mean CDS is obtained by substituting Equation (17) into Equation (15); it is presented in Figure 3. Similar results are also obtained for other values of C_1.

The second size-dependent exit intensity function takes the form

$$\mu_i = c_2/i \qquad (18)$$

where c_2 is a constant. Equation (18) indicates that the exit transition intensity of a crystal is inversely proportional to its size. Thus, a smaller crystal has larger probability of exiting from the system. Often, the term fines-trap or fines-removal is associated with this type of deviation from the idealized crystallizer (2). In this case, Equation (11) becomes

$$p_i(t) = \begin{cases} e^{-(\lambda+c_2)t} & , i=1 \\ \sum_{k=1}^{i}\left\{\left[\dfrac{(\lambda)^{i-1}}{\prod\limits_{\substack{\ell=1\\\ell\neq k}}^{i} c_2(\frac{1}{\ell}-\frac{1}{k})}\right]e^{-(\lambda+\frac{c_2}{k})t}\right\} & , i=2,3,\ldots \end{cases} \qquad (19)$$

The simulated transient mean CSD's under various exit transition intensities given by Equation (18) are shown in Figures 4 and 5. The corresponding transient variances around the mean CSD's are illustrated in Figures 6 and 7.

DISCUSSION

Figure 2 indicates that the size-dependent residence time of the form given by Equation (16) causes a downward curvature of CSD at small crystal sizes. Nevertheless, in a theoretical study of the effect of crystal breakage on CSD, Randolph (15) has postulated that a downward curvature at the small crystal sizes in a semi-logarithmic plot of CSD against the crystal size can be induced by size-dependent, binary breakage of crystals. This has been numerically simulated by solving the steady-state deterministic population balance equation.

Thus, care must be taken in interpreting the experimental data in identifying the cause or causes of the deviation from the idealized crystallizer.

As can be seen from Figures 4 and 5, the size-dependent exit transition intensity, as expressed in Equation (18), gives rise to the upward curvature at small crystal sizes. This is consistent with the conclusion obtained by earlier investigators based on the deterministic population balance approach (1). This deviation of CSD from that of the idealized crystallizer has often been attributed as due to the size-dependent growth rate or growth rate dispersion (8, 9, 10).

Figures 2 through 7 show that the mean CSD and the variance around it are of the same order of magnitude, thereby revealing clearly the stochastic nature of the crystallization phenomenon; obviously, a conventional deterministic model, able to portray only the macroscopic behavior of the crystallizer, fails to predict this phenomenon.

In the present study, the transient behavior of the non-ideal crystallizer under consideration has been determined by analytically solving the ordinary differential equations; in contrast, the conventional or deterministic approach requires solving the governing partial differential equations. Clearly, the former needs much less effort than the latter.

The present stochastic representation can be generalized in a straightforward manner to account for effects of other factors, such as the size-dependent growth rate, degree of supersaturation, and other forms of the size-dependent residence time, on the behavior of a crystallizer. This can be accomplished by appropriately specifying the transition intensities corresponding to the operating conditions.

Constant Exit Transition Intensity

Suppose that the exit transition intensities are size-independent, and all equal to a constant μ, that is,

$$\mu_1 = \mu_2 = \ldots = \mu$$

This corresponds to the case in which the crystallizer is well-mixed (16). Following the same procedure as that employed in deriving Equation (11), Equations (4) and (6) yield the solution

$$p_i(t) = \frac{e^{-(\lambda+\mu)t}(\lambda t)^{i-1}}{(i-1)!}, \quad i=1,2,\ldots \quad (20)$$

The transient mean CSD and the variance around it, corresponding to Equation (20), have been evaluated by substituting Equation (20) into Equations (14) and (15), respectively. A numerical example for such a crystallization process is illustrated in Figures 8 and 9. Again, the magnitude of the mean CSD and the variance around it are of the same order of magnitude, indicating that the prediction of the crystallization dynamics based solely upon the mean behavior of CSD will be untenable; it is essential that stochastic approach be adopted.

Steady-State Solution

The mean CSD and the variance around it under steady-state condition are recovered by letting $t \to \infty$ in the Equations (14) and (15), respectively. Figures 2 through 7 also include the resultant steady-state values for the case when the residence times of crystals are size-dependent. The steady-state mean CSD and the variance around it for the size-independence case are also included in Figures 8 and 9, respectively. It can be seen that the steady-state mean CSD in Figure 8 is linear on a semi-logarithmic plot of the crystal number density against the crystal size. It can be shown that if the residence time of a crystal is size-independent, the steady-state mean CSD reduces to the results of the corresponding deterministic population balance model. Suppose that the rate of seed appearance in the crystallizer is constant, i.e., $z(s)=z$. The steady-state mean CSD can be evaluated by substituting Equation (20) into Equation (14) and letting $t \to \infty$. Since

$$\lim_{t \to \infty} p_i(t) = 0, \quad i=1,2,\ldots$$

we have

$$E[U_i(\infty)]$$

$$= \lim_{t \to \infty} \left[\int_0^t z p_i(t-s)ds + m(0)p_i(t) \right]$$

$$= z \lim_{t\to\infty} \int_0^t \frac{e^{-(\lambda+\mu)(t-s)}[\lambda(t-s)]^{i-1}}{(i-1)!} ds$$

$$= \frac{z}{\lambda}\left(1 + \frac{\mu}{\lambda}\right)^{-i}, \quad i=1,2,\ldots, \tag{21}$$

Thus, the expected total number of crystals inside the crystallizer at steady-state is

$$E[U_t(\infty)] = \sum_{i=1}^{\infty} \frac{z}{\lambda}\left(1 + \frac{\mu}{\lambda}\right)^{-i}$$

and consequently, the fraction of the expected number of crystals in state i becomes

$$\frac{E[U_i(\infty)]}{E[U_t(\infty)]} = \frac{z}{E[U_t(\infty)]\lambda}\left(1 + \frac{\mu}{\lambda}\right)^{-i}, \quad i=1,2,\ldots \tag{22}$$

Let G be the growth rate of a crystal; then, it is clear that

$$G = \lambda\delta \tag{23}$$

Since G has a dimension of length/time, λ has a dimension of 1/time, and δ a dimension of length. Moreover, the exit transition intensity μ is equal to the reciprocal of the mean residence time of the crystallizer, τ. Under the steady-state operation, the number of crystals (seeds) in the feed stream is equal to the number of crystals in the product stream, i.e.,

$$\frac{z}{E[U_t(\infty)]} = \mu = \frac{1}{\tau} \tag{24}$$

and

$$L = i\delta \tag{25}$$

where L denotes the size of a crystal. Thus, Equation (22) reduces to

$$\frac{E[U_i(\infty)]}{E[U_t(\infty)]} = \frac{\delta}{\tau G}\left(1 + \frac{\delta}{\tau G}\right)^{-L/\delta}, \quad i=1,2,\ldots \tag{26}$$

If we let n(L) be the normalized continuous CSD, as $\delta \to 0$, Equation (26) gives

$$n(L)dL = \frac{1}{\tau G} e^{-(L/\tau G)} dL \tag{27}$$

This expresses the fraction of crystals with sizes in the range (L,L+dL); it is in agreement with the steady-state CSD of an idealized MSMPR crystallizer obtained by the deterministic population balance approach (2).

The probability distribution of random variable X can be characterized by its moments. For example, the first moment, the mean or expected value, locates the center of the distribution; the second moment about the mean, the variance, reflects the spread or dispersion of the distribution; the third moment about the mean describes the asymmetry or skewness of the distribution; and the fourth moment about the mean measures the flatness (the kurtosis) of the distribution near its center. The deterministic model is only able to generate the mean CSD and can not provide higher moments around this mean CSD; therefore, it can be considered only as a special case of the present stochastic model (see APPENDIX A). In fact, the mean value of the stochastic model, in general, can be proven to be equivalent to that of the corresponding deterministic population balance model (4, 16).

CONCLUSIONS

The effect of size-dependent residence time on CSD has been modeled stochastically. Depending upon the functional form of the exit transition intensities, various deviations of CSD from that of the idealized crystallizer are predicted. Results of the simulation indicate that a downward curvature at its lower end might be caused by a size-dependent residence time of crystals rather than crystal breakage; an upward curvature of CSD at its lower end could also be generated by the size-dependent residence time of crystals but not necessarily by the size-dependent growth rate or growth rate dispersion.

The present stochastic model is capable of predicting the transient behavior of a crystallizer more readily than the deterministic population balance approach; the steady-state solution of the present stochastic model can also be obtained in a straightforward manner. Higher moments around the mean CSD based on the present model can be evaluated without much effort; this is not the case if the deterministic approach is adopted.

Predicting the dynamics of a crystallizer based on the mean CSD is unreliable since this mean and the variance around it are of the same order of magnitude. Only under special conditions does the stochastic representation reduce to the corresponding deterministic population balance model.

ACKNOWLEDGEMENT

The authors wish to acknowledge Engineering Experiment Station, Kansas State University for partial financial support of this work.

NOTATION

c_1, c_2	constants in the size-dependent exit transition functions
$E[\cdot]$	expectation operator
G	growth rate of a crystal
L	linear crystal size
$m(0)$	number of the inventory seeds in state 1
n^o	number of nuclei
$n(L)$	crystal size distribution
$N_i(t-s)$	number of the crystals in state i in the crystallizer at time t which have come from seeds appearing in the time interval (s, s+ds)
$p_i(t)$	probability that a crystal is in state i at time t
s, t	time scale
$U_i(t)$	cumulative number of crystals in state i in the crystallizer up to time t
$Var[\cdot]$	variance operator
z	rate of seed appearance in the crystallizer

Greek Letters

λ, λ_i	transition intensity
δ	discretizing parameter indicating the size of the state
μ, μ_i	exit transition intensity
τ	mean residence time

LITERATURE CITED

1. Bourne, J.R. and M. Zabelka, Chem. Eng. Sci., 35, 533 (1980).
2. Randolph, A.D. and M.A. Larson, Theory of Particulate Processes, Academic Press, New York (1971).
3. Katz, S. and R. Shinnar, Ind. Eng. Chem., 61(4), 60 (1969).
4. Gardiner, C.W., Handbook of Stochastic Methods for Physics, Chemistry and the Natural Sciences, Springer-Verlag, New York (1983).
5. Randolph, A.D. and M.A. Larson, AIChE J., 8, 639 (1962).
6. Sherwin, M.B., R. Shinnar and S. Katz, AIChE J., 13, 1141 (1967).
7. Wey, J.S. and J. Estrin, Ind. Eng. Chem. Proc. Des. Dev., 12, 236 (1973).
8. Canning, T.F. and A.D. Randolph, AIChE J., 13, 5 (1967).
9. Abegg, C.F., J.D. Stevens, and M.A. Larson, AIChE J., 14, 118 (1968).
10. Berglund, K.A. and M.A. Larson, AIChE J., 30, 280 (1984).
11. Ramanarayanan, K.A., K. Athreya and M.A. Larson, AIChE Symp. Ser., 80 (240), 76 (1984).
12. Larson, M.A., E.T. White, K.A. Ramanarayanan and K.A. Berglund, AIChE J., 31, 90 (1985).
13. Nyvlt, J., O. Sohnel, M. Matuchova and M. Broul, The Kinetics of Industrial Crystallization, Elsevier, New York (1985).
14. Mood, A.M., F.A. Graybill and D.C. Boes, Introduction to Theory of Statistics, McGraw-Hill, New York (1974).
15. Randolph, A.D., Ind. Eng. Chem. Fund., 8, 58 (1969).
16. Nicolis, G. and I. Prigogine, Self-Organization in Nonequilibrium Systems, John Wiley & Sons, New York (1977).
17. Fan, L.T., L.S. Fan and R.F. Nassar, Chem. Eng. Sci., 34, 1172 (1979).

APPENDIX A: MOMENTS OF DISTRIBUTION AND MOMENTS OF RANDOM VARIABLE

To determine the mean transient characteristics of crystallizer through a deterministic population balance approach, in general, partial differential equations need be solved. For simplicity, in lieu of solving such equations, the moments of resultant mean CSD are often defined so that we can obtain partial information on CSD by solving ordinary differential equations of the moments. For a certain CSD, n(L), the n-th moment is defined as (6)

$$r_n = \int_0^\infty L^n n(L) dL, \quad n=0,1,2,... \quad (A-1)$$

It is obvious that given a CSD, moments of any order can be calculated from this equation. The moments are defined by Equation (A-1) is the moments of the deterministic or mean distribution, n(L).

The moments defined in the present study, e.g., Equations (14) and (15) in the text, refer to those of the random variables U_i's. The random variables U_i, i=1,2,..., can take any integer value from 1 to ∞ with certain probabilities; this implies that the number of crystals of size $i\delta$ can be any positive integer but governed by a certain probability law. Furthermore, a certain realization of CSD, $[U_1 = u_1, U_2 = u_2,...]$, based on the present stochastic model, its moments can also be calculated by Equation (A-1); in particular, the moments of the distribution $[U_1 = E(U_1), U_2 = E(U_2), ...]$ correspond to the moments of CSD obtained by the deterministic population balance model. Figure A-1 illustrates the relationship between the mean CSD and the random variables $U_1, U_2,...$ It is worth noting that the first moments of the random variables, $U_1, U_2,...$ as given by Equation (14) in the text comprise collectively the mean CSD as indicated in Figure A-1; obviously it corresponds to CSD generated by the deterministic model.

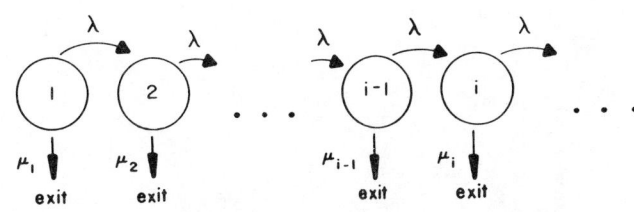

Figure 1. Transition diagram of the crystal growth process for a state-dependent exit transition intensities and constant transition intensities among states.

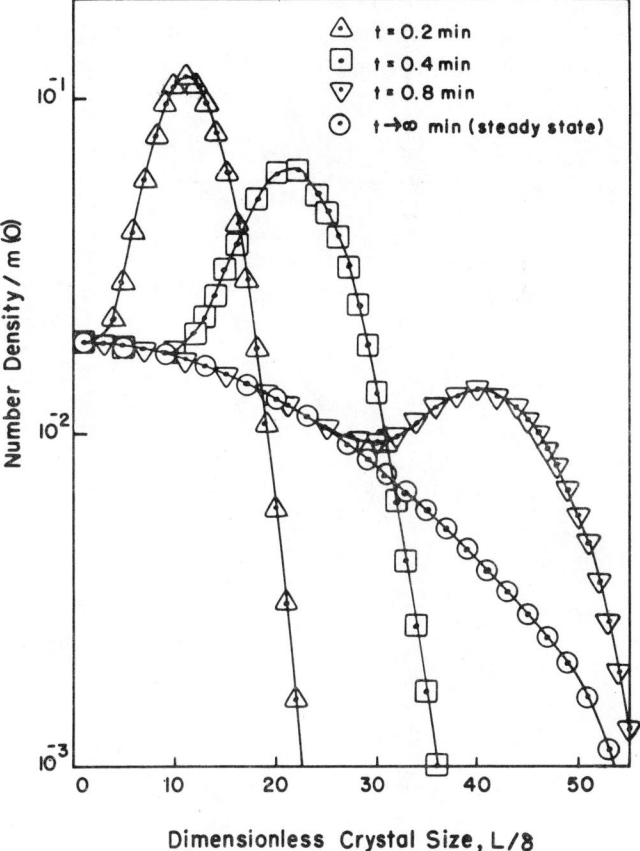

Figure 2. Transient and steady-state mean CSD's for the case of the size-dependent residence time: λ = 54/min, μ_i = 0.1i/min.

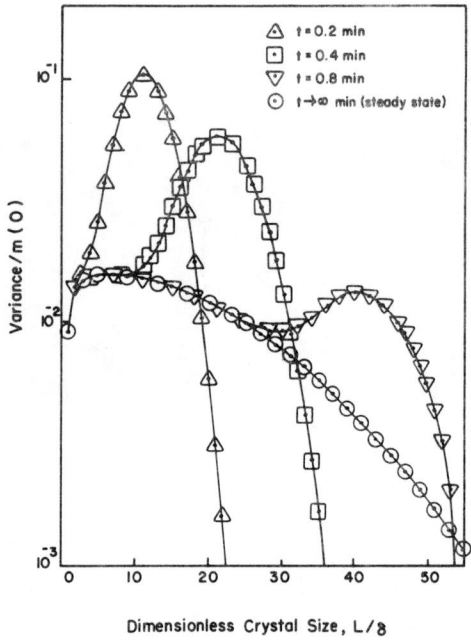

Figure 3. Transient and steady-state variances around the mean CSD's for the case of the size-dependent residence time: $\lambda = 54/\text{min}$, $\mu_i = 0.1i/\text{min}$.

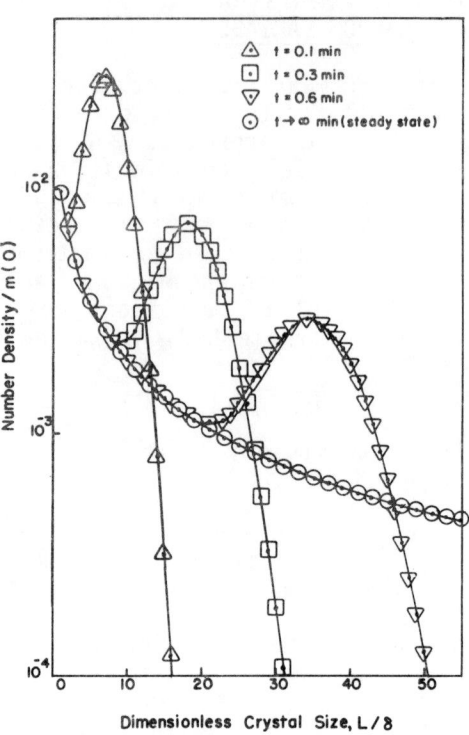

Figure 5. Transient and steady-state mean CSD's for the case of the size-dependent residence time: $\lambda = 54/\text{min}$, $\mu_i = (50/i)/\text{min}$.

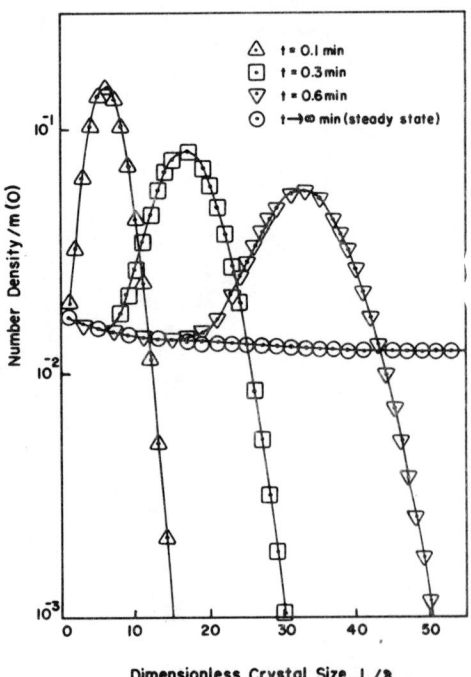

Figure 4. Transient and steady-state mean CSD's for the case of the size-dependent residence time: $\lambda = 54/\text{min}$, $\mu_i = (5/i)/\text{min}$.

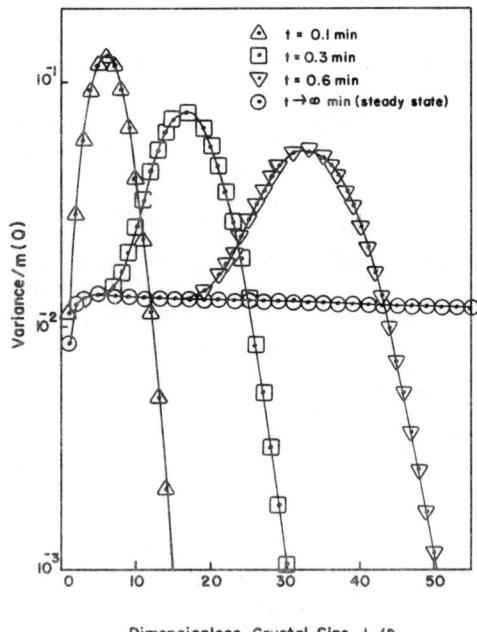

Figure 6. Transient and steady-state variances around the mean CSD's for the case of the size-dependent residence time: $\lambda = 54/\text{min}$, $\mu_i = (5/i)/\text{min}$.

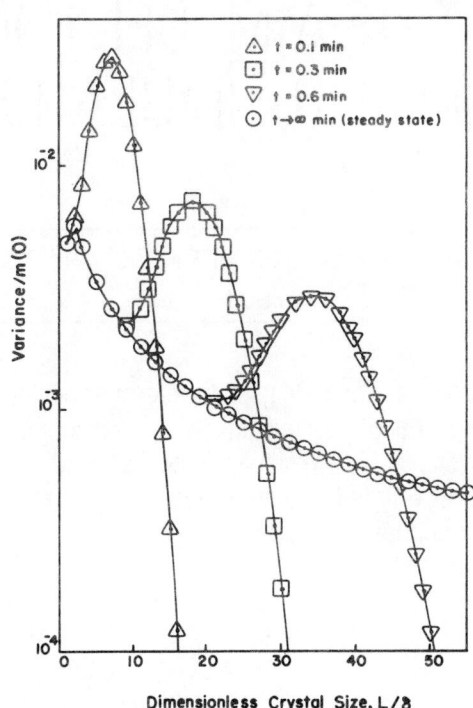

Figure 7. Transient and steady-state variances around the mean CSD's for the case of the size-dependent residence time: $\lambda = 54$/min, $\mu_i = (50/i)$/min.

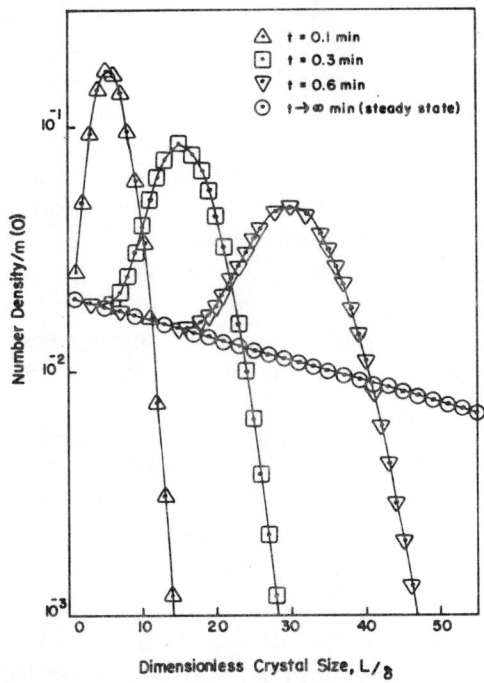

Figure 8. Transient and steady-state mean CSD's for the case of constant growth rate and constant residence time: $\lambda = 50$/min, $\mu_i = 1$/min.

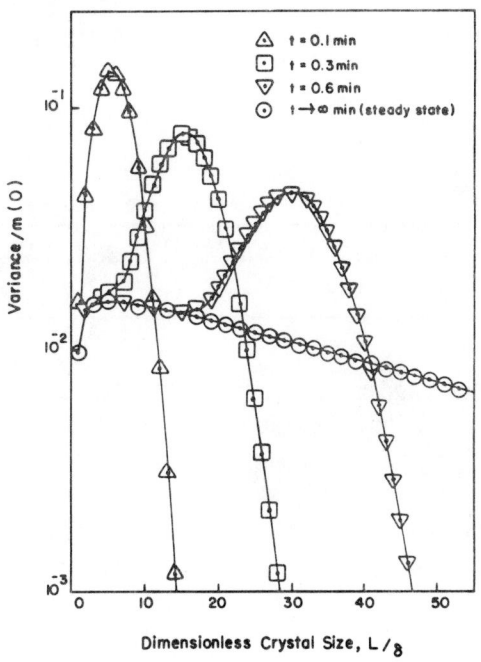

Figure 9. Transient and steady-state variances around the mean CSD's for the case of constant growth rate and constant residence time: $\lambda = 50$/min, $\mu = 1$/min.

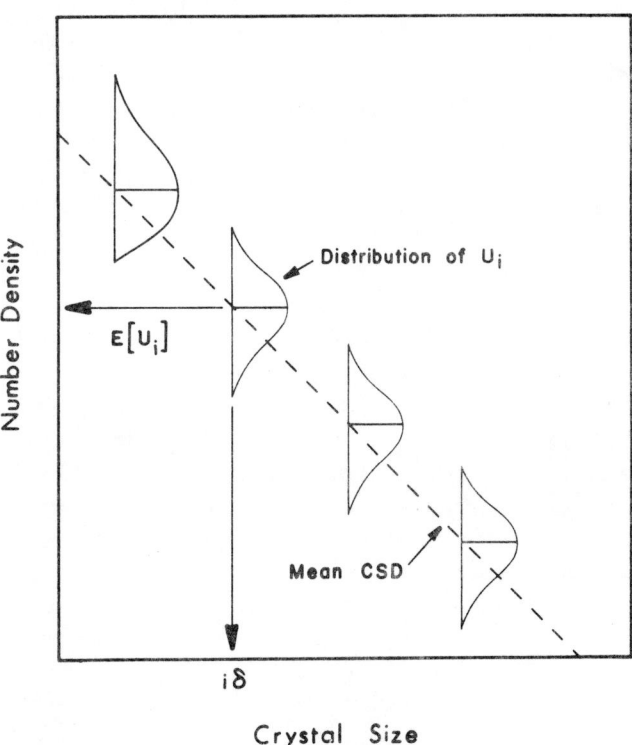

Figure A-1. Relationship between the mean CSD and random variables U_i's.

UTILIZATION OF INDUSTRIAL DATA IN THE DEVELOPMENT OF A MODEL FOR CRYSTALLIZER SIMULATION

Ronald C. Zumstein and Ronald W. Rousseau ■ North Carolina State University, Department of Chemical Engineering, Raleigh, NC 27695-7905

Slurry samples and process operating conditions of a continuous industrial crystallizer were used to determine nucleation kinetics and to develop a mathematical model for crystallization of $CuSO_4 \cdot 5H_2O$. The overall nucleation rate was found to be dependent on magma density, slurry residence time, and power input to the system. Slurry residence time was used in place of crystal growth rate as a measure of supersaturation in the crystallizer and was statistically more significant than growth rate in correlating nucleation kinetics, and nucleation kinetics were highly dependent on agitation.

INTRODUCTION

The crystal size distribution (CSD) in a product from a crystallizer can be determined by simultanously solving population, material, and energy balances which are coupled to nucleation and growth kinetic expressions. Beckman and Randolph (1) and Rousseau and Howell (2) used this concept to simulate continuous complex crystallizers, and to analyze schemes for controlling CSD. In the present study, the model developed by Rousseau and Howell will be modified to describe an industrial copper sulfate pentahydrate ($CuSO_4 \cdot 5H_2O$) crystallizer.

An expression relating process variables to nucleation kinetics is essential to model industrial crystallizers. As suggested in several reviews on the subject (3-7), secondary nucleation is the dominant mechanism of nuclei birth in an industrial crystallizer. Secondary nucleation theory has been used to suggest that supersaturation, concentration of crystals in the crystallizer, and dynamics of the suspension affect the rate of nucleation, and much effort in recent years has been given to scaling up nucleation kinetic expressions from laboratory crystallizers. The recent review by Garside (7) summarizes the work on this phenomena that has produced

---

North Carolina State University, Raleigh, North Carolina. R. W. Rousseau is now with Georgia Institute of Technology, Atlanta, Georgia.

correlations between nucleation rates and design variables. Although these results are promising, a secondary nucleation kinetic expression seems unique to a specific crystallizer.

Several studies (8-13) on secondary nucleation have been conducted using copper sulfate pentahydrate crystals and secondary nucleation kinetic expressions have been developed using fluidized-bed laboratory crystallizers (8-11). The hydrodynamics in these crystallizers make the results of little use in predicting nucleation rates in an industrial crystallizer, but the work does indicate a dependence of nucleation rate on supersaturation and agitation. Fasoli and Conti (12) and Kuboi et al. (13) also have studied $CuSO_4 \cdot 5H_2O$ nucleation rates in a mixed-suspension crystallizer with mechanical agitation. Both studies investigated attrition rates of crystals suspended in a non-solvent. Kuboi et al. (13) indicated that crystal-impeller and crystal-crystal contacts were both important in the production of nuclei. Both studies confirmed the dependence of stirrer speed and magma density on the secondary nucleation rate of copper sulfate.

The purpose of the present study is to demonstrate how a mathematical model can be developed for a continuous, industrial crystallizer. The work concentrates on the determination of the nucleation kinetic expression from industrial data, and demonstrates the type of data that are required for model development, how such data can be obtained, and what model assumptions should be tested in data analysis. In addition, the resulting kinetic expression is used to demonstrate the utility of the model in predicting the effect of several process variables on CSD.

FUNDAMENTAL RELATIONSHIPS

Randolph and Larson (14) have shown that a population balance on a perfectly-mixed (mixed-suspension mixed-product removal, or MSMPR) crystallizer at steady state leads to a single population density function. If all crystal sizes have the same mean residence time τ and all crystals have the same growth rate G, the population density function is given by

$$n(L) = n° \exp(-\frac{L}{G\tau}) \quad (1)$$

where $n°$ is the effective nuclei population density and L is a characteristic dimension defining crystal size. As shown by Saeman (15), the differential weight distribution w for the MSMPR crystallizer is a gamma distribution function:

$$w(L) = \frac{1}{6G\tau}(\frac{L}{G\tau})^3 \exp(-\frac{L}{G\tau}) \quad (2)$$

The cumulative mass distribution for the MSMPR crystallizer, determined by integrating Equation 2 from 0 to L, is

$$W(L) = 1 - \left[1 + \frac{L}{G\tau} + \frac{(\frac{L}{G\tau})^2}{2} + \frac{(\frac{L}{G\tau})^3}{6}\right] \exp(-\frac{L}{G\tau}) \quad (3)$$

The nucleation rate $B°$ is related to the nuclei population density $n°$ by

$$B° = n° G \quad (4)$$

Since the exact form of the kinetic expression for secondary nucleation is unknown, a power-law expression of the following form usually is used to correlate nucleation kinetics:

$$B° = k_N \sigma^n M_T^j \quad (5)$$

where σ is the relative supersaturation and the parameter k_N is a function of temperature, impurity concentration, and crystallizer hydrodynamics.

Crystal growth rates are also usually related to crystallizing conditions by a power-law expression of the form

$$G = k_g \sigma^m \quad (6)$$

where k_g is a function of temperature and impurity concentration. Equation 6 can be substituted into Equation 5 to eliminate supersaturation to give

$$B° = k_N G^i M_T^j \quad (7)$$

Population densities can be estimated from sieve data by the expression

$$n(L) = \frac{\Delta M}{V_s \rho k_V (\bar{L})^3 \Delta L} \quad (8)$$

where $\bar{L}$ is the average crystal size for crystals left on a sieve, ΔM is the mass of crystals left on a sieve, V_s is the volume of the suspension sample, and ΔL is the size difference between the largest and smallest crystals on a sieve. An arithmetic mean of the two adjacent screen sizes usually is used for $\bar{L}$.

THE INDUSTRIAL CRYSTALLIZER

The crystallizer used in the present work was a 22,620-gal (85,630-L) vacuum Struthers-Wells unit used to produce copper sulfate pentahydrate. The product is produced in several grades, determined by size, and ranged in mean size from 0.25 mm to 2.5 mm. As the product fraction that corresponds to each of the grades is important to the commercial success of the operation, it is extremely important that the dependence of CSD on operating conditions be known. Such knowledge can then be used as the basis for producing desired quantities of each product grade.

A schematic of the crystallizer is shown in Figure 1. The recirculation rate through the barometric leg is controlled by a variable-speed elbow pump. The suspension, through visual observations, appears well-mixed below the recirculation withdrawal line. Above this point, the suspension is not as turbulent and few crystals are present. The level of the suspension is maintained by an overflow at the top of the chamber. The overflow was observed to be virtually free of crystals (less than 2% by weight) during normal operation due to the quiescent region above the recirculation withdrawal line. The crystallizer is equipped with fines destruction capabilities that are used in manufacturing certain product grades. (In the present work, however, data from fines destruction operations were not used.) The well-insulated crystallizer utilizes

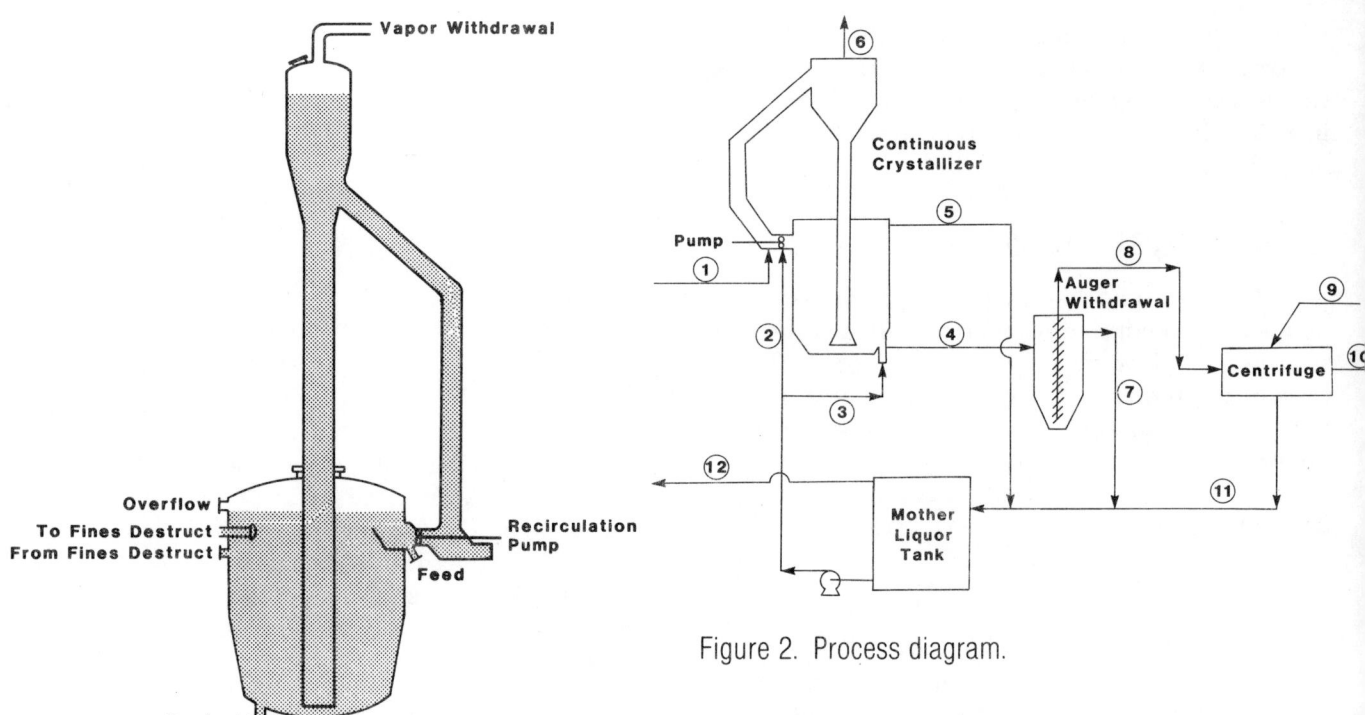

Figure 1. Continuous vacuum crystallizer.

Figure 2. Process diagram.

TABLE 1. Range of Operating Conditions

Feed Rate	15-25 gpm
Return Rate	10-50 gpm
Product Rate	10-50 gpm
Recirculation Rate	4800-6100 gpm
Overflow Rate	20-60 gpm
Vapor Rate	1.8-4.5 lb_m/min
Chamber Temperature	78-85 °F
Return Temperature	82-112 °F
Feed Temperature	152-190 °F
Pump Current	81-119 A
Magma Density	0.334-0.639 g/mL

only evaporation to remove the sensible heat of the feed streams.

Figure 2 shows the process flow diagram for the system without fines destruction. Hot, concentrated copper sulfate solution 1 and a return stream 2 of saturated solution, both of which are free of crystals, are fed to the crystallizer at the suction of the recirculation pump. The return stream is necessary to maintain the level in the crystallizer. The product slurry 4 is removed from the crystallizer, often with the aid of a carrier stream 3 of mother liquor. The flow rate of water vapor is represented by 6 and the overflow by 5, which is discharged into the mother-liquor tank. Typical operating conditions are shown in Table 1.

Only the operating data on the crystallizer were available for the study; however, the operation of other units in Figure 2 can be described in terms of visual observations. The product slurry 4 from the crystallizer is fed into an auger, where the solids content of the product stream 8 is increased. Liquor with a very low solids content is also removed from the auger and sent to the mother-liquor tank 7. The product stream 8 is sent to a centrifuge where the crystals 10 are removed for washing and drying. Wash water 9 is added to the centrifuge and the liquor 11 is removed. Heat is added to 11 so that fines present in 7 and 11 will dissolve in the mother-liquor tank. This heat addition maintains the mother-liquor tank at temperatures 10 to 20 °F greater than the crystallizer. During normal operation the mother-liquor tank contains copper sulfate solution very near saturation; it also has some crystalline material, but this is only on the bottom of the tank. Overflow from the mother-liquor tank 12 is returned to another part of the plant.

DATA FROM THE CRYSTALLIZER

The data were obtained on an irregular basis from the copper sulfate crystallizer and they consisted of temperatures, feed concentrations, flow rates, and sieve analyses. The feed flow rate 1 was determined using an existing flow meter while all other flow rates were measured with a sonic flow meter. Sieve analyses were performed on crystals recovered from 725-mL samples of slurry removed from the crystallizer chamber at two locations: one-third and two-thirds down from the top of the crystallizer. Also, a sieve analysis was performed on a sample of the product crystals from the centrifuge. Slurry samples were taken with a depth sampler that was lowered from the top of the crystallizer. Seventeen sets of data were collected during production without fines removal.

The first step in the determination of the nucleation kinetic expression was to estimate the mean residence time, crystal growth rate, and effective nuclei density at the operating conditions of each run. The mean residence time of the crystallizer was determined from the product flow rate:

$$\tau = \frac{V}{Q_p} \quad (9)$$

where V is the crystallizer suspension volume (22 620 gal) and Q_p is the product flow rate (4 minus 3). As Randolph and Larson (14) suggested for MSMPR crystallizers, population densities estimated from sieve data by Equation 8 can be fit to Equation 1 and used to determine the effective nuclei density $n°$ and the crystal growth rate G for each run.

Before the sieve data can be transformed into population densities by Equation 8, the volume shape factor k_V, relating the crystal characteristic dimension L to the crystal volume, had to be determined. Small samples of crystals were sieved into several fractions, each with a known characteristic dimension. A group of crystals from each fraction was then counted and weighed, and the volume shape factor was determined by the relationship

$$k_V = \frac{W}{\rho \overline{L}^3 \Delta N} \quad (10)$$

where ΔN is the number of crystals in the sample of mass W. As shown in Figure 3, the equivalent density ρk_V was determined to be 2.23 g/mL and independent of crystal size. Since the density of copper sulfate pentahydrate is 2.286 g/mL, the

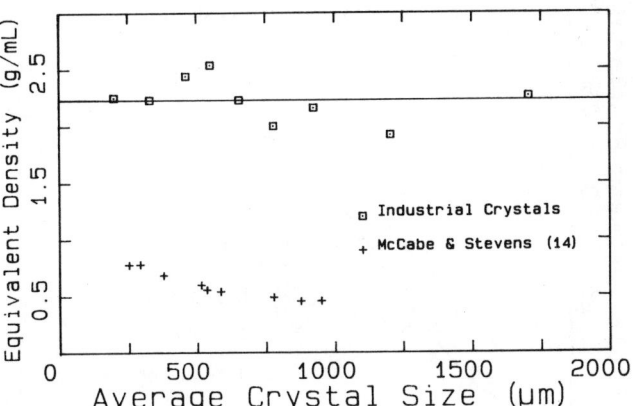

Figure 3. Determination of volume shape factor.

volume shape factor is approximately 0.975. Figure 3 also presents the values of equivalent density reported by McCabe and Stevens (16). The copper sulfate pentahydrate crystals used by McCabe and Stevens to determine the equivalent density were produced from solutions of higher purity than those producing the industrial crystals. Impurities in solutions often alter the habit of growing crystals, which can be quantitatively described by changes in the crystal shape factor. The solutions in the industrial opeation contain an excess of sulfuric acid and impurities such as iron and nickel, which also have been reported to alter the habit of some crystalline substances.

Typical sieve data for two slurry samples and a product sample are shown in the cumulative mass fraction distribution of Figure 4. The curve represents a non-linear regression fit of the two slurry samples to Equation 3. The crystal size distributions of the two samples from the crystallizer are in good agreement, indicating that the crystallizer is well mixed. The percentages of the smaller crystals in the product are less than the corresponding samples from the crystallizer magma, confirming the earlier suggestion that some of the fine crystals were lost from the magma in flowing through the auger and centrifuge.

Population densities were determined from the sieve data by Equation 8, and the values from two crystallizer slurry samples were averaged to produce one set of population densities for each run. Values of $n°$ and G were determined from all 17 sets of data by a linear least-squares fit of Equation 1 to the estimated population densities. Equation 1

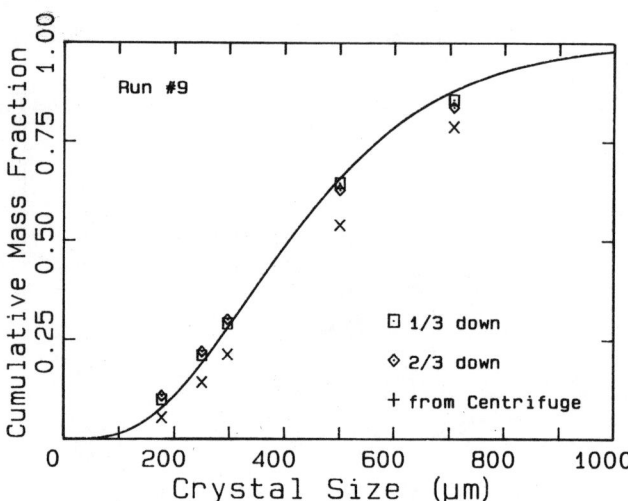

Figure 4. Typical cumulative mass distribution data from slurry samples.

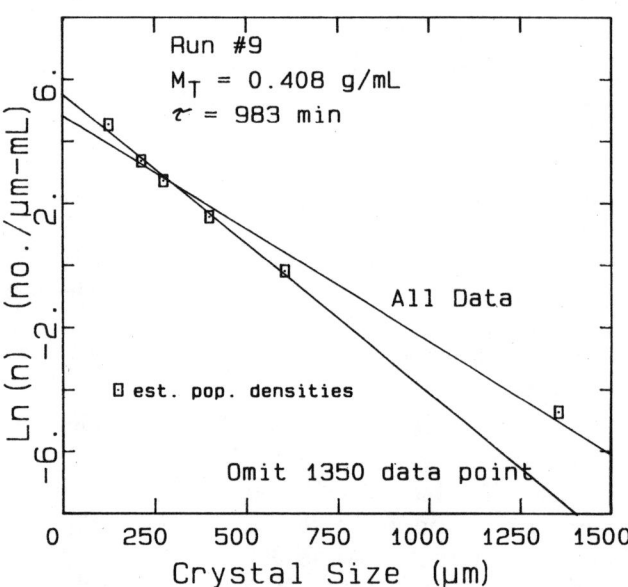

Figure 5. Typical population density data.

compared with the average experimental values obtained from the weight of crystals in each 725-mL sample. The average difference between the measured and calculated magma densities was 33%. This large difference is a measure of the poor fit of Equation 1 to all the data.

As indicated by the data in Figure 5, the population density correlation is weighted heavily by the value at approximately 1350 μm. Moreover, the sieve nest used to size the crystals had an inappropriate spread between the 10- and 25-mesh (2000 and 707 μm) screens. Anywhere from 3 to 30% of the total crystal mass was left on the 25-mesh screen, and usually none was left on the 10-mesh screen. Thus, the average size of the crystals left on the 25-mesh screen was probably less than the arithmetic average of the screen sizes. Consequently, new values of $n°$ and G were determined from the data after deletion of population density data at 1350 μm, and Equation 11 was used to obtain new values of magma densities for each of the data sets. The average difference between the measured magma densities and those determined with the new model parameters was only 4.4%, a considerable improvement over the original correlation. The effect of fitting measured population densities after deleting the 1350-μm data is shown in Figure 5.

It is clear from Figure 5 that the population density data are in excellent agreement with the population density function given by Equation 1, indicating that the system functions as an MSMPR crystallizer. Also, no abnormalities such as size-dependent growth, growth rate dispersion, or internal size classification were evident in any of the 17 data sets. The overflow, which may be assumed to carry a small portion of the fines from the crystallizer, does not appear to cause a large deviation from the assumption of MSMPR operation. (Such a deviation would result in a change of slope in the population density function at small size.

The above approach for determining $n°$ and G for an MSMPR crystallizer has used population densities which have been estimated from sieve data. The population densities estimated by Equation 8 are dependent on both the range of sizes collected on a sieve and the estimate of the average crystal size on a sieve. The population densities at 1350 μm were omitted because the large size range between the 10- and 25-mesh screens (1293 μm) caused a poor estimate of the average size. Consequently, the above method for determining $n°$ and

with the correlated values of $n°$ and G is plotted along with the estimated population densities in Figure 5 for a typical set of data. It is quite apparent that the population density function given by Equation 1 is a poor fit to all the data. However, the measured magma density must be consistent with that calculated from the values of $n°$ and G:

$$M_T = 6k_V \rho n°(G\tau)^4 \qquad (11)$$

Magma densities calculated using the values of $n°$ obtained by correlating all sieve data were then

G from sieve data can only be used to approximate these parameters due to the inherent error in estimating population densities. This problem can be overcome by first using Equation 3 to determine the growth rate and then using Equation 11 to determine nucleation rate. Equation 3 gives the cumulative weight fraction of crystals left on a screen of size L. Discrete points of this function are easily obtained from raw sieve data without approximations. A non-linear regression fit of the sieve data to Equation 3 can then be performed to determine the parameter $G\tau$. The growth rate is then determined from knowledge of the product flow rate and Equation 9. The nucleation rate can be determined directly from the reported value of the magma density by rearranging Equation 11 to give

$$B^\circ = \frac{M_T}{6\rho k_V \tau (G\tau)^3} \quad (12)$$

The results obtained by using the procedure of the preceding paragraph to analyze the 17 data sets are given in Table 2, and a comparison of the population density function with that obtained from fitting population density data is shown in Figure 6. The use of cumulative weight fractions is thought to provide better estimates of the kinetic parameters in an MSMPR crystallizer than the method involving population densities. Before using this method, however, the population densities should first be used to check the validity of assumptions that the crystallizer operates as an ideal MSMPR crystallizer and no growth abnormalities occur.

KINETIC EXPRESSIONS

Since no measurement of chamber concentration was made, a growth rate kinetic expression could not be determined from the industrial data. However, Toyokura et al. (5) reported that the growth rate of copper sulfate pentahydrate at 85 °F followed the expression

$$G = 7970 (\Delta C)^{1.6} \quad (13)$$

where ΔC is the supersaturation expressed in terms of g $CuSO_4 \cdot 5H_2O$/g H_2O. This correlation was obtained by fitting data from experiments in which growth rates ranged from 2.4 to 24 μm/min. It is important to point out that the data on which Equation 13 is based were obtained on solutions free from impurities, which is quite different from the present situation. However, impurities normally influence the magnitude of the growth rate and have less influence on the order relating growth rate and supersaturation.

TABLE 2. Estimated Nucleation Kinetic Parameters Using Equations 3 and 12

Run No.	Average M_T (g/mL)	Estimated $G\tau$ (μm)	Reported τ (min)	Calculated B° (no./μm·mL)
8	0.452	114	443	51.1
9	0.408	111	984	22.5
10	0.334	144	1140	7.35
11	0.517	98	2060	19.7
12	0.436	89	901	51.1
13	0.396	103	637	42.5
14	0.345	128	571	21.4
15	0.457	89	637	77.4
16	0.499	97	610	67.2
17	0.381	159	884	7.97
18	0.386	110	1140	19.3
19	0.639	88	635	110.
20	0.442	118	1240	16.3
21	0.532	101	983	39.0
22	0.380	94	1060	32.3
23	0.511	102	1030	34.9
24	0.547	105	1010	35.4

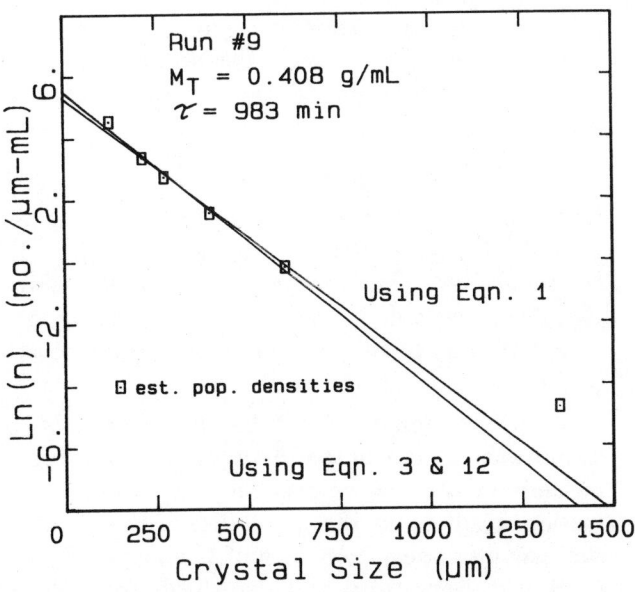

Figure 6. Comparison of correlation from fit of population density data with that from fit of cumulative mass distributions.

A nucleation kinetic expression for the industrial copper sulfate crystallizer can be obtained from the MSMPR data by fitting the data in Table 2 to Equation 8. In addition to changes in growth rate and magma density between runs, hydrodynamic changes also occurred due to changes in the recirculation rate. The current to the recirculation pump was monitored constantly during operation and was assumed to control magma hydrodynamics. Since the voltage across the pump motor was constant, the measurement of current could be converted to power input to the crystal suspension. Adding the pump current A (in amperes) as an independent variable to Equation 8, a multivariable least-squares regression on the kinetic data gave the nucleation expression

$$B° = 1.4 \times 10^{-11} G^{0.2} M_T^{0.4} A^{6.4} \quad (14)$$

with a correlation coefficient of 0.76. The variables M_T and A were significant at the 81% and 99% confidence levels, respectively, while the probability that G was a significant variable in the expression was only 18%.

As suggested by Grootscholten et al. (18), overall slurry residence time rather than growth rate can be used as the measure of the solution concentration. They used pilot-plant data from sodium chloride crystallization to determine that secondary nucleation rates were a function of magma density, residence time, and agitation. As residence time increases, the concentration of the liquor should approach the saturation concentration due to an extended period of growth. Performing the regression with τ instead of G, the following expression was obtained:

$$B° = 1.9 \times 10^{-6} M_T^{0.7} \tau^{-0.7} A^{4.8} \quad (15)$$

with a correlation coefficient of 0.82. The variables τ and A were both significant at the 95% confidence level, but the probability that M_T was a significant variable in the expression was only 55%. The actual fit of this nucleation kinetic expression to the data is presented in Figure 7.

The decrease in probability that M_T is significant is attributed here either to errors in measurement or to the existence of a solids concentration gradient in the crystallizer. Based on five successive samples from each of two locations in the crystallizer (one-third and two-thirds from the top of the crystallizer), it was determined that the reproducibility in determinations of magma density was between 5 and 10%. However, there was an

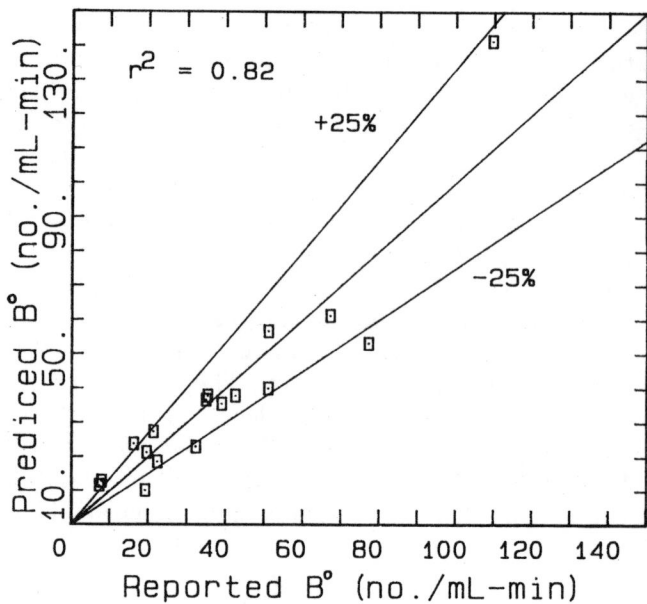

Figure 7. Comparison of correlation with experimentally determined nucleation rates.

average of 16% difference in the average magma densities from the two locations, despite the fact that the sieve analyses for the two slurry samples were in excellent agreement.

CRYSTALLIZER MODEL

As stated earlier, the objective of this study was to develop a model that can be used to relate measured operating variables to crystal size distribution. The development of the model follows the work of Rousseau and Howell (2) and produces a modified version of their algorithm called GEM. An overall material balance around the crystallizer was added to GEM so that a clear-liquor overflow rate could be calculated. The program first determines the steady-state crystal size distribution based on the measured operating variables by simultaneously solving the steady-state balance equations and the kinetic expressions. The solvent vapor flow rate from top of the barometric leg was first estimated by a steady-state energy balance. The three other balance equations—population, overall mass, and solute—along with the two kinetic expressions were used to determine the chamber concentration, overflow rate, growth rate, nucleation rate, and magma density. The solution procedure utilies two *regula-falsi* loops to iterate to the solution. Once the steady-state crystal size distribution has been determined, disturbances in the operating conditions can be input and a dynamic simulation performed. The three balance equations, containing appropriate accumulation terms, were again solved simultaneously with the kinetic expressions. The solution procedure utilizes a finite difference approximation for the population balance and a Newton-Raphson iteration scheme. As in GEM, the model can be used for a crystallizer with fines destruction and product classification.

The model was used first to calculate the steady-state crystal size distributions corresponding to operating conditions reported for the 17 crystallizer runs that had been used to develop an expression for nucleation kinetics. Agreement between measured and predicted crystal size distributions for nine of the data was excellent, as is illustrated for one of these runs in Figure 8. The lack of agreement in the remaining eight data sets is thought to be due to errors in the measured crystallizer conditions; flow rate measurements were accurate to within ±10%, and the assumption of saturation of the return stream could be invalid under certain circumstances. Manipulation of the measured operating variables by less than 10% could be used to bring the model predictions and experimental data into the sort of agreement illustrated in Figure 8.

The model relies on the validity of the two kinetic expressions, one for nucleation and the other for growth. The expression for nucleation kinetics was developed from data on the operating system, while that for growth is based on data taken on relatively pure solutions at 85 °F. The conditions in the present system were such that the crystallization temperature varied between 80 and 115 °F, and the liquor contained impurities such as iron, nickel, and sulfuric acid. Both of these deviations in conditions affect the parameter k_G in the growth kinetic expression. For example, a large value of k_G in Equation 5 causes supersaturation to be low in the crystallizer, which corresponds to Class II behavior. Table 3 presents results obtained by using different values of k_G to obtain the predicted steady-state CSD. The last case in Table 3 assumes that the concentration in the chamber is at saturation so that inclusion of a growth rate expression in the model is unnecessary. These results indicate that for a Class II, the steady-state predictions are not highly dependent on the validity of the growth kinetic expression and a good approximation of the CSD can be determined without a growth rate expression.

TABLE 3. Effect of Growth Expression on CSD

Expression	ΔC	σ	n^o	G
$G = 7920.(\Delta C)^{1.6}$	0.00097	0.00236	155	0.121
$G = 792.0(\Delta C)^{1.6}$	0.00410	0.01000	154	0.121
$G = 79.20(\Delta C)^{1.6}$	0.01725	0.04207	151	0.120
None	0.00000	0.00000	155	0.121

CONCLUSIONS

The following conclusions have resulted from the application of crystallization theory to the industrial system examined in the present work:

1. The operation of the industrial crystallizer shown in Figure 1 can be approximated by an ideal MSMPR model. The small concentration of solids in the overflow causes little deviation in the CSD from the model. No internal size classification of crystals is evident, but a concentration gradient of solids appears to exist in the upper portion of the crystallizer.

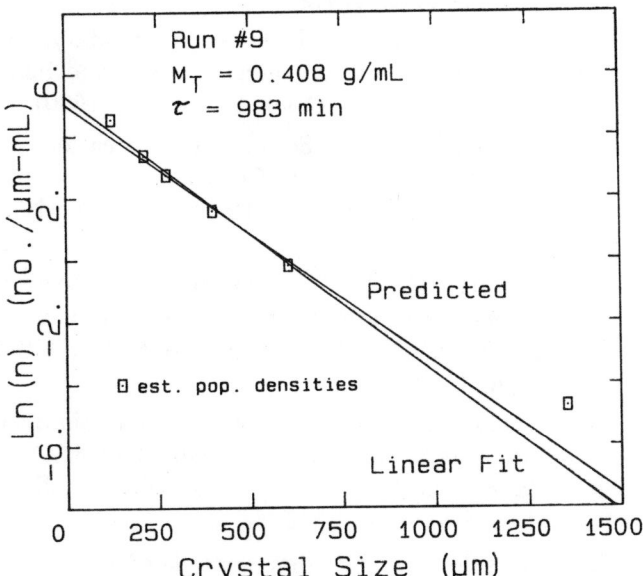

Figure 8. Typical comparison of predicted and measured population densities.

2. The secondary nucleation of copper sulfate pentahydrate in this particular industrial crystallizer fits the kinetic expression given by Equation 15. The expression indicates that the nucleation rate is highly dependent on the degree of agitation. Nucleation theory (7) and laboratory studies with copper sulfate (11) both indicate that the nucleation rate should be dependent on the degree of agitation to the 3.0 to 4.5 power, which is only slightly less than observed. Magma residence time was a more significant variable in the kinetic expression than was growth rate. Although both variables can be used to represent the dependence of nucleation rate on supersaturation, it appears that residence time has an additional effect on nucleation rate which makes it a more significant variable.

3. If a system behaves as an MSMPR crystallizer, Equations 3 and 12 should be used to determine nucleation and growth rates. Due to the approximation of population densities from sieve data, determination of nucleation and growth rates by fitting Equation 1 is less accurate the the proposed method.

4. The mathematical model is able to predict adequately the steady-state CSD in the crystallizer when accurate estimates of input variables are known. Errors in the reported input variables of as little as 10% cause a very poor estimate of the CSD. The model is not dependent on the validity of the growth kinetic expression for Class II systems.

NOMENCLATURE

- A — current to recirculation pump (A)
- $B°$ — nucleation rate (no./mL·min)
- G — linear crystal growth rate (μm/min)
- k_g — growth rate kinetic constant in Equation 6
- k_N — constant in Equation 5
- k_N' — constant in Equation 13
- k_V — crystal volume shape factor (dimensionless)
- L — characteristic one-dimenional crystal size (μm)
- $\bar{L}$ — arithmetric average of two adjacent sieve sizes (μm)
- ΔL — difference between two adjacent sieve sizes (μm)
- ΔM — mass of crystals left on sieve (g)
- M_T — magma density (g crystals/mL slurry)
- ΔN — number of crystals (no.)
- $n(L)$ — population density at size L (no./μm·mL)
- $n°$ — effective nuclei density (no./μm·mL)
- Q_p — product slurry flow rate (gpm or L/min)
- ρ — crystal density (g/mL)
- σ — relative supersaturation (dimensionless)
- τ — suspension residence time (min)
- V — crystallizer volume (gal or L)
- V_s — suspension sample volume (mL)
- $W(L)$ — weight fraction of crystals less than L (dimensionless)
- $w(L)$ — mass distribution function of crystals (1/μm)

LITERATURE CITED

1. Beckman, J. R. and A. D. Randolph, "Crystal Size Distribution Dynamics in a Classified Crystallizer: Part II. Simulated Control of Crystal Size Distribution," *AIChE J.*, **23**, 510, (1977).

2. Rousseau, R. W. and T. R. Howell, "Comparison of Simulated Crystal Size Distribution Control Systems Based on Nuclei and Supersaturation," *Ind. Eng. Chem. Process Des. Dev.*, **21**, 606, (1982).

3. Garside, J. and R. J. Davey, "Secondary Contact Nucleation: Kinetics, Growth and Scale-up," *Chem. Eng. Commun.*, **4**, 393 (1980).

4. Botsaris, G. D., "Secondary Nucleation - A Review," *Industrial Crystallization*, J. W. Mullin, eds., Plenum Press, New York, 3, 1976.

5. deJong, E. J., "Nucleation - A Review," *Industrial Crystallization 78*, S. J. Jancic and E. J. deJong, eds., North-Holland, Amsterdam, 3, 1979.

6. Larson, M. A., "Secondary Nucleation: An Analysis," *Industrial Crystallization 81*, S. J. Jancic and E. J. deJong, eds., North-Holland, Amsterdam, 55, 1982.

7. Garside, J., "Review Article Number 15: Industrial Crystallization from Solutions," *Chem. Eng. Science*, **40**, 3 (1985).

8. Karpinski, P. H., "Crystallization as a Mass Transfer Phenomenon," *Chem. Eng. Science,* **35**, 2321 (1980).

9. Karpinski, P. H. and K. Toyokura, "Secondary Nucleation and Growth of Copper Sulphate Crystals in a Fluidized Bed," *Industrial Crystallization 78,* S. J. Jancic and E. J. deJong, eds., North-Holland, Amsterdam, 55, 1979.

10. Karpinski, P. H., "Effect of Hydrodynamics on Secondary Nucleation and Growth Rates in a Fluidized Bed," *Industrial Crystallization 81,* S. J. Jancic and E. J. deJong, eds., North-Holland, Amsterdam, 97, 1982.

11. Toyokura, K., M. Uchiyana, M.Kawai, H. Akutsu, and T. Ueno, "Secondary Nucleation of $KAl(SO_4)_2 \cdot 12H_2O$, $MgSO_4 \cdot 7H_2O$ and $CuSO_4 \cdot 5H_2O$," *Industrial Crystallization 81,* S. J. Jancic and E. J. deJong, eds., North-Holland, Amsterdam, 87, 1982.

12. Fasoli, U. and R. Conti, "Crystal Breakage in a Mixed Suspension Crystallizer," *Kristall und Technik,* **8**, 931 (1973).

13. Kuboi, R., A. W. Nienow, and R. Conti, "Mechanical Attrition of Crystals in Stirred Vessels," *Industrial Crystallization 84,* S. J. Jancic and E. J. deJong, eds., Elsevier Science, 211 (1984).

14. Randolph, A. D. and M. A. Larson, *Theory of Particulate Processes,* Academic Press, New York, 1971.

15. Saeman, W. C., "Crystal-Size Distributions in Mixed-Suspension," *AIChE J.,* **2**, 107 (1956).

16. McCabe, W. L. and R. P. Stevens, "Rate of Growth of Crystals in Aqueous Solutions," *Chem. Eng. Prog.,* **47(4)**, 168 (1951).

17. Grootscholten, P. A. M., L. D. v. d. Brekel, and E. J. de Jong, "Effect of Scale-up on Secondary Nucleation Kinetics for the Sodium Chloride-Water System," *Chem. Eng. Res. Des.,* **62**, 179 (1984).

INDEX

A

Adsorption 8
Agglomeration 78,85
Ammonium dihydrogen phosphate 47
Analysis 120
Aqueous solutions 1

B

Barium sulfate 85
Batch 110
BCF Surface diffusion theory 54
Bimodal CSD 85

C

Calcium carbonate precipitation 62
Calcium sulfate 23
Coarsening 95
Contact nucleation 19
Contaminant/impurity 62
Control 110
Cooling crystallizers 42
Copper sulfate 130
Coprecipitation/adsorption 62
Crystal growth 8,36,54,85
Crystal growth models 54
Crystallite aging 95
Crystallization 23,54,78,110
Crystallization rate 42
Crystallizer 130
Crystallizer dynamics 104
Crystallizer modeling 104
Crystal size distribution 104,130
CSD 120
CSD control 104

D

Diesel fuel 31
Dissolution 36

F

Feedback 110

G

Growth model 31
Gypsum 23

I

Impurity effect 19
Inhibition 36
Ionic solutions 8
Isomorphic inclusion 62

K

Kinetics 23,36,78
Kinetics of crystal growth 54

L

Lead (Pb) 62

M

Magnesium sulfate heptahydrate 47
Mechanisms 8
MSMPR 78,120
Multiparticle diffusion 95

N

Nonuniform conditions 47
Nucleation kinetics 130

O

Optimization of sugar crystallization 42
Ostwald ripening 95

P

Particle size distribution 62
Population balance 85

Potassium sulfate 78
Properties 1

S

Secondary nucleation 19
Simulation 110
Size distribution 23
Solution growth 47
Solutions, supersaturated 1
State space formulations 104
Stochastic 120
Strontium fluoride 36

Sucrose 42
Supersaturated 1
Surface reaction 8
Survival steps 19
Synthetic products 31

T

Transient 120

W

Wax 31

SYMPOSIUM SERIES

ADSORPTION

- 96 Developments in Physical Adsorption
- 117 Adsorption Technology
- 219 Recent Advances in Adsorption and Ion Exchange
- 230 Adsorption and Ion Exchange—'83
- 233 Adsorption and Ion Exchange—Progress and Future Prospects
- 242 Adsorption and Ion Exchange: Recent Developments

AEROSPACE

- 33 Rocket and Missile Technology
- 52 Chemical Engineering Techniques in Aerospace

BIOENGINEERING

- 69 Bioengineering and Food Processing
- 84 The Artificial Kidney
- 86 Bioengineering ... Food
- 93 Engineering of Unconventional Protein Production
- 99 Mass Transfer in Biological Systems
- 108 Food and Bioengineering—Fundamental and Industrial Aspects
- 114 Advances in Bioengineering
- 163 Water Removal Processes: Drying and Concentration of Foods and Other Materials
- 172 Food, pharmaceutical and bioengineering—1976/77
- 181 Biochemical Engineering Renewable Sources of Energy and Chemical Feedstocks

CRYOGENICS

- 224 Cryogenic Processes and Equipment 1982
- 251 Cryogenic Properties, Processes and Applications 1986

CRYSTALLIZATION

- 110 Factors Influencing Size Distribution
- 193 Design Control and Analysis of Crystallization Processes
- 215 Nucleation, Growth and Impurity Effects in Crystallization Process Engineering
- 240 Advances in Crystallization From Solutions
- 253 Fundametnal Aspects of Crystallization and Precipitation Processes

DRAG REDUCTION

- 111 Drag Reduction
- 130 Drag Reduction in Polymer Solutions

ENERGY

Conversion and Transfer

- 5   Heat Transfer, Atlantic City
- 57 Heat Transfer, Boston
- 59 Heat Transfer; Cleveland
- 75 Energy Conversion Systems
- 79 Heat Transfer with Phase Change
- 87 Advances in Cryogenic Heat Transfer
- 113 Convective and Interfacial Heat Transfer
- 118 Heat Transfer—Tulsa
- 119 Commercial Power Generation
- 138 Heat Transfer—Research and Design
- 162 Energy and Resource Recovery from Industrial and Municipal Solid Wastes
- 174 Heat Transfer: Research and Application
- 189 Heat Transfer—San Diego 1979
- 202 Transport with Chemical Reactions
- 208 Heat Transfer—Milwaukee 1981
- 216 Processing of Energy and Metallic Minerals
- 225 Heat Transfer—Seattle 1983
- 236 Heat Transfer—Niagara Falls 1984

Nuclear Engineering

- 53 Part XIII
- 56 Part XIV
- 94 Part XX
- 104 Part XXI
- 106 Part XXII
- 119 Commercial Power Generation
- 168 Heat Transfer in Thermonuclear Power Systems
- 169 Developments in Uranium Enrichment
- 191 Nuclear Engineering Questions Power Reprocessing, Waste, Decontamination Fusion
- 221 Recent Developments in Uranium Enrichment

ENVIRONMENT

- 78 Water Reuse
- 97 Water—1969
- 115 Important Chemical Reactions in Air Pollution Control
- 122 Chemical Engineering Applications of Solid Waste Treatment
- 124 Water—1971
- 126 Air Pollution and its Control
- 133 Forest Products and the Environment
- 137 Recent Advances in Air Pollution Control
- 139 Advances In Processing and Utilization of Forest Products
- 144 Water—1974: I. Industrial Wastewater Treatment
- 145 Water—1974: II. Municipal Wastewater Treatment
- 146 Forest Product Residuals
- 147 Air: I. Pollution Control and Clean Energy
- 148 Air: II. Control of NO_{xx} and SO_x Emissions
- 149 Trace Contaminants in the Environment
- 151 Water—1975
- 156 Air Pollution Control and Clean Energy
- 157 New Horizons for the Chemical Engineer in Pulp and Paper Technology
- 165 Dispersion and Control of Atmospheric Emissions, New-Energy-Source Pollution Potential
- 170 Intermaterials Competition in the Management of Shrinking Resources
- 171 What the Filterman Needs to Know About Filtration
- 175 Control and Dispersion of Air Pollutants: Emphasis on NO_X and Particulate Emissions
- 177 Energy and Environmental Concerns in the Forest Products Industry
- 184 Advances in the Utilization and Processing of Forest Products
- 188 Control of Emissions from Stationary Combustion Sources Pollutant Detection and Behavior in the Atmosphere
- 195 The Role of Chemical Engineering in Utilizing the Nation's Forest Resources
- 196 Implications of the Clean Air Amendments of 1977 and of Energy Considerations for Air Pollution Control
- 198 Fundamentals and Applications of Solar Energy
- 200 New Process Alternatives in the Forest Products Industries
- 201 Emission Control from Stationary Power Sources: Technical, Economic and Environmental Assessments
- 207 The Use and Processing of Renewable Resources—Chemical Engineering Challenge of the Future
- 209 Water—1980
- 210 Fundamentals and Applications of Solar Energy II
- 211 Research Trends in Air Pollution Control: Scrubbing, Hot Gas Clean-up, Sampling and Analysis
- 213 Three Mile Island Cleanup
- 223 Advances in Production of Forest Products
- 232 Applications of Chemical Engineering in the Forest Products Industry
- 239 The Impact of Energy and Environmental Concerns on Chemical Engineering in the Forest Products Industry
- 243 Separation of Heavy Metals and Other Trace Contaminants
- 246 Advances in Process Analysis and Development in the Forest Products Industries.

FLUIDIZATION

- 101 Fundamental Processes in Fluidized Beds
- 105 Fluidization Fundamentals and Application
- 116 Fluidization: Fundamental Studies Solid-Fluid Reactions, and Applications
- 176 Fluidization Application to Coal Conversion Processes
- 205 Recent Advances in Fluidization and Fluid-Particle Systems
- 234 Fluidization and Fluid Particle Systems: Theories and Applications
- 241 Fluidization and Fluid Particle Systems: Recent Advances

HISTORY OF CHEMICAL ENGINEERING

The History of Penicillin Production | 235 Diamond Jubilee Historical/Review Volume

ION EXCHANGE

Adsorption and Ion Exchange Separations | 230 Adsorption and Ion Exchange—'83 | 233 Adsorption and Ion Exchange—Progress and Future Prospects
Recent Advances in Adsorption and Ion Exchange

KINETICS

Reaction Kinetics and Unit Operations | 73 Kinetics and Catalysis

MINERALS

Mineral Engineering Techniques | 173 Fundamental Aspects of Hydrometallurgical Processes | 180 Spinning Wire from Molten Metals
Fossil Hydrocarbon and Mineral Processing | | 216 Processing of Energy and Metallic Minerals

PETROCHEMICALS

Polymer Processing | 135 The Petroleum/Petrochemical Industry and the Ecological Challenge | 142 Optimum Use of World Petroleum
Declining Domestic Reserves—Effect on Petroleum and Petrochemical Industry | | 212 Interfacial Phenomena in Enhanced Oil Recovery

PETROLEUM PROCESSING

C_4 Hydrocarbon Production and Distribution | 135 The Petroleum/Petrochemical Industry and the Ecological Challenge | 155 Oil Shale and Tar Sands
Declining Domestic Reserves—Effect on Petroleum and Petrochemical Industry | 142 Optimum Use of World Petroleum | 226 Underground Coal Gasification: The State of the Art

PHASE EQUILIBRIA

Pittsburgh and Houston | 6 Collected Research Papers | 88 Phase Equilibria and Gas Mixtures Properties

PROCESS DYNAMICS

Process Dynamics and Control | 55 Process Control and Applied Mathematics | 214 Selected Topics on Computer-Aided Process Design and Analysis
Process Systems Engineering

SEPARATION

Recent Advances in Separation Techniques | 192 Recent Advances in Separation Techniques—II | 250 Recent Advances in Separation Techniques—III

SONICS

109 Sonochemical Engineering

TRANSPORT PROPERTIES

56 Selected Topics in Transport Phenomena

MISCELLANEOUS

Chemical Engineering Reviews | 194 Hazardous Chemical—Spills and Waterborne Transportation | 231 Data Base Implementation and Application
Small-Scale Equipment for Chemical Engineering Laboratories | 203 A Review of AIChE's Design Institute for Physical Property Data (DIPPR) and Worldwide Affiliated Activities | 237 Awareness of Information Sources
Engineering, Chemistry, and Use of Plasma Reactors | | 238 New Developments in Liquid-Liquid Extractors: Selected Papers From ISEC '83
Vacuum Technology at Low Temperatures | 204 Tutorial Lectures in Electrochemical Engineering and Technology | 244 Experimental Results from the Design Institute for Physical Property Data. I: Phase Equilibria
Standardization of Catalyst Test Methods | 206 Controlled Release Systems | 247 Chemical Engineering Data Sources
Biorheology | 217 New Composite Materials and Technology | 248 Industrial Membrane Processes
The Modern Undergraduate Laboratory Innovative Techniques | 220 Uncertainty Analysis for Engineers | 249 Measurement of High Temperatures in Furnaces and Processes
Electro Organic Synthesis Technology | 228 Problem Solving |
Plasma Chemical Processing | 229 Tutorial Lectures in Electrochemical Engineering and Technology—II | 252 Thin Liquid Film Phenomena
Chronic Replacement of Kidney Function

MONOGRAPH SERIES

The Manufacture of Nitric Acid by the Oxidation of Ammonia—The DuPont Pressure Process by Thomas H. Chilton | 7 The 'Calculated' Loss-of-Coolant Accident by L.J. Ybarrondo, C.W. Solbrig, H.S. Isbin | 11 Lumps, Models and Kinetics in Practice by Vern W. Weekman, Jr.
Experiences and Experiments with Process Dynamics by Joel O. Hougen | 8 Understanding and Conceiving Chemical Process by C. Judson King | 12 Lectures in Atmospheric Chemistry by John H. Seinfeld
Present, Past, and Future Property Estimation Techniques by Robert C. Reid | 9 Ecosystem Technology: Theory and Practice by Aaron J. Teller | 13 Advanced Process Engineering by James R. Fair
Catalysts and Reactors by James Wei | 10 Fundamentals of Fire and Explosion by Daniel R. Stull | 14 Synfuels from Coal by Bernard S. Lee
| | 15 Computer Modeling of Chemical Processes by J.D. Seader
| | 16 "High-Tech" Materials by Sheldon Isakoff

RAYMOND H. FOGLER LIBRARY
DATE DUE

BOOKS ARE SUBJECT TO
RECALL AFTER TWO WEEKS

ISBN 0-8169-0425-1